Metropolis and Province

Metropolis and Province,

Science in British Culture, 1780–1850

Ian Inkster and
Jack Morrell, Editors

University of Pennsylvania Press
Philadelphia
1983

ISBN 0-8122-7855-0
Library of Congress Card Number   82-40354

Printed in Great Britain

Contents

List of Tables		6
List of contributors		7
Preface		9

1 Introduction: Aspects of the history of science and science culture in Britain, 1780–1850 and beyond
Ian Inkster — 11

2 Whigs and savants: reflections on the reform movement in the Royal Society, 1830–48
Roy M. MacLeod — 55

3 The London lecturing empire, 1800–50
J. N. Hays — 91

4 The British Mineralogical Society: a case study in science and social improvement
Paul Weindling — 120

5 'Nibbling at the teats of science': Edinburgh and the diffusion of science in the 1830s
Steven Shapin — 151

6 Science in a commercial city: Bristol 1820–60
Michael Neve — 179

7 Rational dissent and provincial science: William Turner and the Newcastle Literary and Philosophical Society
Derek Orange — 205

8 Economic and ornamental geology: the Geological and Polytechnic Society of the West Riding of Yorkshire, 1837–53
Jack Morrell — 231

9 Medical elites, the general practitioner and patient
     power in Britain during the cholera epidemic of 1831–2
     Michael Durey   257

     General index   279

     Name index   284

Tables

1 Profile of the British Mineralogical Society,
   1799–*c.* 1806   138
2 Attendance at the Mineralogical Society   141
3 Profile of the Askesian Society, 1796–*c.* 1808   143
4 Topics and number of papers presented to the YGS,
   1838–53   249
5 Topics and authors of papers presented to the YGS,
   1838–53   249

Contributors

Michael Durey is Lecturer in History, School of Social Inquiry, Murdoch University, Perth, Western Australia, 6150, Australia. Past interests: history of cholera in Britain, and the pre-famine Irish in Britain. Current interests: demography of Perth, Western Australia, 1860–1920, and the origins of the Provincial Medical and Surgical Association.

J. N. Hays is Associate Professor of History, Department of History, Loyola University of Chicago, 6525 N. Sheridan Road, Chicago, Illinois, 60626, USA. Past interests: Victorian popular science and scientific institutions. Current interests: the historical geography of industrialization in Great Britain and the USA.

Ian Inkster is Senior Lecturer, Department of Economic History, University of New South Wales, PO Box 1, Kensington, NSW, 2033, Australia. Past interests: social history of science, education and medicine in England prior to 1914, and general economic history of technology. Current interests: social history of science in England, economic history of technology in Japan, and role of science and technology in underdeveloped economies.

Roy M. MacLeod is Professor of Modern History, Department of History, University of Sydney, Sydney, New South Wales, Australia. Past and current interests: social history of science, medicine and technology, the history of science education, and the historical sociology of the learned professions.

Jack Morrell is Lecturer, School of Social Sciences, University of Bradford, Bradford, West Yorkshire, BD7 1DP. Past and current interests: British science during the first industrial revolution.

Michael Neve is Lecturer, Unit of History of Medicine, Department of Anatomy, University College London, Gower Street, London, WC1E 6BT. Past interests: social history of nineteenth-century science, especially in the West Country. Current interests: history of psychiatry in Britain in the nineteenth century; science in London, 1790–1870.

Derek Orange is Senior Lecturer, Department of Humanities, Newcastle Polytechnic, Ellison Place, Newcastle upon Tyne, NE1 8ST. Past interests: history of nineteenth-century chemistry. Current interests: provincial science and education in nineteenth-century Britain; the historical relations of science and religion.

Steven Shapin is Lecturer, Science Studies Unit, Edinburgh University, 34 Buccleuch Place, Edinburgh, EH8 9JT, Scotland. Past interests: Scottish science, social history of British science in the eighteenth and nineteenth centuries, popular education in science. Current interests: sociology of scientific knowledge, seventeenth- and eighteenth-century natural philosophy, social uses of scientific knowledge.

Paul Weindling is Research Assistant, Wellcome Unit for History of Medicine, University of Oxford, 47 Banbury Road, Oxford, OX2 6PE. Past interests: Oscar Hertwig, relations between cell theory and Darwinism, secularization in late eighteenth-century Europe. Current interests: the development of scientific organizations in Britain 1780–1820; biology, medicine and society in nineteenth- and twentieth-century Germany; bibliographical resources in history of science and medicine.

Preface

This book focuses on British scientific culture during the first industrial revolution. We intend it to be useful to university and college level history students, and to be a guide for post-graduates in history of science who are seeking methods and materials. We also hope that it will interest all those who, from their varying perspectives, appreciate the importance of cultural formations.

The book reflects recent changes which have taken place in historians' approaches to the study of past science. As an intended contribution to social history of science, it looks at science and scientific culture in their local contexts, whether metropolitan or provincial, examining where possible the relations between metropolitan and provincial science and savants.

The very title of the book has determined the order in which the chapters appear. Inkster's introduction attempts to draw general conclusions from the various essays, while also providing a new analytical framework for understanding the popularity of scientific culture in Britain during her early industrialization and urbanization.

The opening three essays cover aspects of science in London. MacLeod examines changes in the select Royal Society of London – Europe's oldest continuing scientific society – in relation to the general triumph of political reform. Hays looks at that Society's cultural hinterland, while Weindling deals with the institutionalization of specialized interests. The rest of the essays, while not ignoring relations with London, are primarily concerned with the provinces. Shapin explores the paradox of Edinburgh's position as a provincial metropolis. In contrast, Neve and Orange demonstrate the differences in the nature of scientific culture between two provincial settings. Such variation was associated with differing notions of utility, which Morrell explores through the case of geology in a mining area. Finally Durey deals with a period of crisis in which scientific knowledge was subject to specifically local pressures which in turn were conditioned by provincial–metropolitan relations.

We are grateful to the contributors for their forbearance with us. In their advice and help, Claire L'Enfant, Emily Wheeler, and their colleagues at Hutchinson have exceeded the call of duty. We also owe a debt to Lorraine Inkster and Janet Morrell for the sane and sublime indifference which they have consistently shown to this joint enterprise.

Ian Inkster, Sydney
Jack Morrell, Bradford

1 Introduction: Aspects of the history of science and science culture in Britain, 1780–1850 and beyond

Ian Inkster

I

> In the field of ideas the age of revolutions gave little of importance comparable with that of the scientific discoveries or technical inventions of the period. Time was needed to digest the events and transformations that followed each other in rapid succession from 1760 to 1830.
>
> J. D. Bernal, *Science in History*, 1969, ii, p.542.

If to Bernal's generalization we add Ashton's observation that during the industrial revolution in Britain, 'there was much coming and going between the laboratory and the workshop', and our knowledge of the contemporaneous popularity of scientific culture, then there would appear to be some justification for a specifically "economic" approach to the history of science in Britain during the years approximately 1780–1850.[1]* Indeed, until comparatively recent years, the historiographical trend has been to explain the diffusion of science in just such terms. The present essay suggests that a social history of British scientific culture and institutions is more appropriate. Taken together, the essays in the present volume expose the popularity of science in its several forms, over very varied locations and on widely differing levels. Furthermore, although explanations vary because of the difference in data examined, the essays do clearly converge on the basis of a social history of scientific culture.

Although many in number and large in volume, the attempts to forge the direct *causal* link between science and economy, or, more succinctly, between scientific interests and industrial technology, have failed to satisfy economic historians and have neglected a range of historical data.[2] They do, however, serve to demonstrate convincingly the enormous popularity of science at all levels

---

*Superior figures refer to the Notes and references which appear at the end of chapters.

(although not *among* all people or all groups) of British society, and, particularly, to highlight the scientific interests of businessmen, technicians, artisans and the less skilled among the working class.[3] Even here, however, difficulties arise. The Manchester Literary and Philosophical Society may be treated as a forum for science and industry and its activity and vigour may thereby be expressed in "industrial utility" terms. But it may also be given, as we shall see in greater detail below, a predominantly "social" treatment and explanation, where the activities of individuals are interpreted in the light of broad social utilities rather than in terms of direct economic interests.[4] Similarly, the science of the Mechanics' Institutes may be seen as an expression of either the economic prerequisites of the nation at large or as a reflection of the economic interests of bourgeoisie or workers. It follows that their seeming "failure" as purveyors of science by the end of our period may be taken as a sign that such economic functions were not being fulfilled or were accomplished more successfully by other competing institutions.[5] Alternatively, the Institutes may be seen as associational offshoots of a radical urban scientific culture or as a focus for cultural domination of one maturing class over another. It follows that their "failure" may then be considered illusory or, if yet acknowledged, attributed to a decline of necessary social rather than economic functions.[6]

Not only may the approach through direct industrial utility often be restated in terms of social factors, but the social approach, as witnessed in the present volume, can incorporate the material of the former approach within its basic argument. For such reasons one of the leading exponents of the "industrial utility" school has readily admitted to the 'immense importance of ... socio-psychological forces, and of the associated educational-scientific-technological factors, in the Industrial Revolution, which cannot, therefore, any longer be explained simply in economic terms of supply and demand'.[7] None of which is to argue that millwrights were not often well versed in mathematics, or that manufacturers did not at times imbibe relevant knowledge of chemical processes in scientific societies and groups, or that much of the appeal at the very popular level was utilitarian and technical in nature.[8] Rather one is saying two things. First, that such instances do not adequately *explain* the rise, nature and progress of cultural and institutional forms. Second, that much material falls outside the triangle of science, technology and industry. The present volume details both these aspects. The possible reply to such points by defenders of a more "economic" approach is that social institutions and the motives of actors within them are themselves determined by economic factors. The reply is not an answer. For although it is to an extent true that social locations are economically conditioned, such a generalization ignores the vital causal and mediating functions of socio-cultural forms, implies the existence of either a more developed class structure prior to the 1840s than in fact emerged (see below, p.139) or a greater dependence of technology on science than in fact characterized the British industrial revolution, and not least dismisses much rich historical material.

The posing of several questions may serve to illuminate the significance of the last point. Why was the scientific culture of Nottingham and Norwich during the period 1790–1830 seemingly so less diffused, institutionalized and vigorous than that of, say, York, Liverpool or Derby?[9] If York's comparatively sophisticated culture was a hangover from the pre-industrial social and economic structure, then what of Norwich? Even if Norwich could be reasonably explained in "economic" terms, is not the cultural-lag effect worthy of examination in its own right? Why did the metropolis as well as the provinces demonstrate an enormous growth of science culture at both the very popular and the informed levels? If businessmen were involved in chemistry, why also, and quite as noteworthy, were they concerned with geology, botany and astronomy? Similarly, why did Unitarian ministers delve deeply into the mysteries of the steam engine? What explains the explosion of the so-called pseudo-sciences, such as phrenology, animal electricity and mesmerism? In such times of trouble as 1796, 1817 or 1832, is it necessarily true that public laudation of the industrial utility of science coincided with the motives of its supporters? More generally, are public rationalizations and justifications the reflection of individual reasons and motives? Why was astronomy possibly the most popular science of the 1790s and 1820s? And why did its proponents and supporters submerge its undoubted potential economic function among such other appeals as entertainment, moral uplift and pure intellectual stimulus? The "industrial utility" approach to scientific culture fails even to ask such questions.

II

Much of the more recent social history of British science in this period accords with Barñes's plea that historians should forsake the widest of generalizations when examining science or science culture:

A better approach is to regard ideas as tools with which social groups may seek to achieve their purposes in particular situations. Ideas *suit* purposes not because of any logical relationship, but because they are naturally suited to particular kinds of use within an existing system of beliefs and norms.[10]

If social systems undergo basic change, then so too do our, and the actors', perceptions of science and the scientific enterprise. Despite important and interesting differences in interpretation, (see Section IV below and Chapters 4 and 6), most recent social history of British science utilizes a non-rigid definition of its subject matter, and attempts to locate scientific interests, activity and, to some extent, creativity, within fairly particular social contexts. Thus the otherwise varied essays in the present volume all recognize that science has social functions in the social system and social uses for the individuals comprising it, that scientists have career interests as well as ideas, that scientific training and research require patronage either via a specific clientele (government, industry, gentlemen amateurs) or a more diffused "audience", and that scientific research programmes may emerge from networks of institutions and individuals. Similarly, they recognize that a wider "scientific movement" (activists) and audience served a mediating function between the scientific community and the society as a whole, and that prior to professionalization this movement may have channelled individuals into science proper and to an extent have served as a training ground.

The approach, common to most recent social history of science, allows a great deal of variety of focus. Through his intellectual mobility John Bostock of Liverpool directly linked the scientific movement with the scientists. On the other hand, William Turner of Newcastle (see Chapter 7) acted as a figurehead linking the "movement" with the wider scientific audience and with local society generally. The Royal Institution of London (founded 1799) may have owed much to the efforts of individuals in coming to terms with the technological changes that were transforming agriculture and industry, and its continuing support may have been even more indebted to the needs for social mobility of specific social groups in the metropolis. In addition, it certainly appears to have 'served to stimulate provincial as well as metropolitan societies'.[11] But it was also the meeting and training ground for such foremost members (or, more significantly, yet-to-be foremost members) of the scientific community as Humphry Davy, Thomas Young, Frederick Accum, W. T. Brande and Thomas Garnett.[12] In a quite different cultural-geographical setting, the Manchester Literary and Philosophical Society (founded 1781) may have been primarily the institutional expression of the social needs of characteristic social types within the scientific movement, but it also served a fundamental role in the development of the "scientific career" of such intellectual leaders as John Dalton, Thomas Henry, Thomas Percival, William Sturgeon and J. P. Joule.[13] The smaller Askesian Society of London (founded 1796) was similarly a focus for social and intellectual gatherings among a very specific social group in the metropolitan setting, but nevertheless it provided a platform for joint research projects, mobility through the scientific community, for example, to fellowship of the Royal Society, and a beginning to the intellectual careers of such members of the British scientific community as Astley Cooper, Richard and William Phillips, W. H. Pepys, Luke Howard, Alexander Tilloch and Wilson Lowry.[14] Such findings are strongly supported by studies which focus on the individual scientific worker and work outwards to the institutional and cultural infrastructure. Thus Michael Faraday's "education" in science incorporated such informal, diffused influences.[15] More recently it has been shown that the geologist Charles Lyell's career between the early 1820s and the early 1840s was decidedly influenced by his participation in the activities of such associations as the coterie of the Geological Society (founded 1807) and its Geological Society Club, the more provincial British Association (founded 1831) and the Royal Institution. Such associational activity was not simply part of Lyell's social "background" but rather a vital ingredient in his personal foreground, as institutions were utilized by him quite purposefully and with calculation for the furtherance of his own scientific career as a devotee.[16]

Because of the highly complex nature of the data, and because the material is only yet being explored there is less academic accord over the question of the motives and social characteristics of individuals within the larger scientific enterprise. If the activities of scientists such as Lyell or Davy may be reasonably explained in terms of career profiles, what of the much greater number of individuals in metropolis or province who comprised the scientific movement and audience, namely, those whose existence converted 'natural and experimental philosophy' into a popular culture?[17] The essays in the present volume, (particularly these of Shapin, Orange, Neve and Morrell) address themselves to precisely this sort of question.

It should be emphasized that this is not to imply that "social" explanations focused on the scientific movement do not also expose the motives and social dependencies of the creative scientists. The opposite may be maintained. We have seen that, as the culture generally provided a "foreground" for the sustenance of creative scientists and, as such scientists emerged from the scientific movement, so too the social characteristics of the one were likely to be related to those of the other. It is true enough, however, that any communality may have been distorted by the overlaying of career and other motives upon the more general motives and characteristics of the scientists. Also, as Thackray has emphasized, the *attraction* of science, and thus the motives *for* science among many individuals within the scientific movement lay in the images of science generated by the scientists themselves (either overtly or through their work), and this, to some extent, was in turn a result of *their* real or apparent social characteristics.[18] Thus Joseph Priestley projected a very definite image of science and of himself, kept alive for many years by the leading members of the scientific community, as perfectly demonstrated in the extremely well publicized Priestley Commemorations of 1833.[19]

One of the most general arguments which might be offered as an explanation for Britain's vital scientific culture during these years is that related to the pervading "individualism" of her scientists and their institutions; this is at least a partial result of state neglect of both science and industry (see Section IV below). If entrepreneurs of the industrial revolution, 'while not enjoying state sponsored and formalised technical training, frequently availed themselves of local opportunities which were usually sufficient for their immediate purposes', then it might further be argued that it was precisely this functional element which buoyed the scientific culture. Given the nature of Britain's industrial revolution, the underlying scientific culture and scientific movement served 'science' well enough.[20] The strength of the scientific culture was a result of an individual and local "initiative" which was itself integral to the whole style of industrialization and urbanization in Britain. Although this argument, if developed, would be attractive and useful because of its emphasis on local variations in initiative and activity, it would also, as stated in the present form, be naively "functional" in its structure and neglectful of the actual motives and characteristics of those taking initiatives. Not all provincials were absorbed into the scientific enterprise. What were the characteristics of those who were?

In a critical essay review of Musson and Robinson's *Science and Technology in the Industrial Revolution*, Berman suggested that,

We need to examine the role of science in the rise of the middle class. We need more studies of the class basis of modern science, empirically grounded on hard statistics of membership and overlapping membership in scientific and related societies. We need to examine the social repercussions of the professionalisation and organisation of science. . . .[21]

Berman himself has since gone someway towards satisfying such needs. Borrowing from Gramsci's concept of 'hegemony' as 'a kind of cultural supremacy which sanctions the social authority of the ruling class', Berman went on to argue that the cultivation of science and scientific research in Britain during these years becomes understandable only when seen within the framework of the 'cultural imprint of the ruling class'.[22] This argument concerning the wide appeal of science is dependent on the assertion that science was *the* cultural child of the ruling class. Any independent, middle-class scientific movement was soon subverted. As Berman puts it:

provincial societies and industrially oriented science offered more than economic rewards; they offered the possibility of transcending class barriers . . . because of the enormous psychological power of the hegemonic class, Baconian science, which could have quickly become a strictly middle class, alternative ideology, was instead vitiated for a long time by the gentleman amateur tradition. . .[of specialist and provincial science]. . . interest in the subjects themselves was secondary to the interest in the social mobility they could afford. Whatever their intellectual value, their social function probably lay in providing easier access to ruling class circles.[23]

So, for Berman, the provincial science which is so fully documented throughout the present volume, was essentially 'the expression of an elite in industrial towns aspiring to more fashionable circles. . .'.[24] This position differs from the essential core of Thackray's conclusions from his study of Manchester. Adopting a group and locale approach much nearer to that recommended by

Barnes (see above, p. 14), Thackray argues that the science activists and audience *were* significantly representative of an alternative value system to that of the ruling class, and that their ultimate success as urbanites was surely an element in the social transformation of British society in these years:

A key to understanding may lie in the social legitimisation of marginal men. Such legitimisation is itself a complex, subtle thing. The adoption of science as a mode of cultural self-expression also depends on a particular affinity between progressivist rationalist images of scientific knowledge and the alternative value system espoused by a group peripheral to English society. . . . The alliance between science, peripheral status, and progressivist philosophy was itself transmuted as the larger culture within which the alliance had formed experienced its own shifts and changes.[25]

Thackray's individuals are much more socially *assertive* than Berman's hegemonized pre-Smilesian social climbers. The attribute both types have in common is, to a greater or lesser extent, their successful upward social mobility. Where they profoundly differ is in the *motivations* attributed to their varied social actions, including the scientific. In cultivating the sciences in province or metropolis, Berman's individuals are seeking an entrance while Thackray's are both representing and seeking an alternative. Without at this stage critically evaluating either approach at the ultimate level of the underlying nature of British society at that time – a subject left for consideration at a later point (Section IV below) – it is instructive to understand the differences between them.

Berman is explicitly writing in class terms, while Thackray is considering an urban group (although his argument is often on grounds which imply an underlying class model). Thus Berman may affirm that a ruling class wielded science and that all those who were engaged in science at any level and who were not of the ruling class, were, therefore, motivated by the needs of upward mobility. Thackray, because of his material, is ready to acknowledge that *not all* of Manchester's educated men were science activists and to mould this recognition into his overall explanation. What was *particular* about those *within* the urban setting who cultivated science, assuming that all those outside the prevailing ruling elite were not simply non-ruling and thus homogeneous in essence? Noting the radicalism and dissent of his group, Thackray goes on to argue that:

Their espousal of the progressivist values of Unitarianism and a progressivist interpretative of science can then be seen as deriving from their need to justify themselves, and to do so in terms of belief systems that simultaneously affirmed their commitments to high culture, announced their distance from the traditional value systems of English society, and offered a coherent explanatory scheme for the unprecedented, change-oriented society in which they found themselves unavoidably if willingly cast in leading roles.[26]

From Thackray's emphasis on characteristic groups it follows that he is also prone to more exact quantitative confirmation of his general claims, heeding Peter Laslett's recent reminder that 'all historians deal in societies to some extent, even if their chosen subjects are individuals or states of mind, and therefore all historians deal in quantities'.[27] Thus his finding that 16 per cent of his 1781 group and 30 per cent of his 1809–11 group of cultivators of science were Unitarian, at a time when that sect represented a tiny minority of even the industrial urban population, carries more authority as regards their particularity, than does Berman's unspecified implications of homogeneity in social actions and of social types.[28]

A third difference is methodological. Berman does appear to argue backwards from results to intentions. Because many of the middle class were ultimately established among the ruling elite, then their motives and the motives of those who sought to emulate them must have been to that end. Thackray allows that motives and perceptions were far more complex, as is illustrated in his "transmutation" of the alliance between science and status. Furthermore, in Thackray's argument the social internalization of individuals within the scientific movement is not solely a result of their own activity and upward mobility, but is also a function of shifts in social institutions which take place throughout the period.

The fourth explanation of difference concerns timing. While Berman's argument is merely pre-1840, Thackray's treatment of social location and motivation is sensitive to chronology. This lends enormous strength to the Thackray approach, allowing a treatment of trends in the scientific movement and culture, for example, the extreme popularity of provincial and metropolitan science in the 1790s and late 1820s to early 1830s. Greater recognition of specific temporal changes in scientific culture would undoubtedly have forced Berman to a more complex formulation. For instance, although the 1840s may illustrate the "social mobility" appeal of natural scnce, surely this would not be true of, say, the mid 1790s or 1816–19, when science was seen as "French philosophy" or otherwise subversive? Thompson and others have shown the *extreme* radical imagery attached to science in the first period.[29] The impact of Sidmouth's Acts of 1817 similarly illustrates the political fragility of science for the later period.[30] More generally, how does Berman's approach accommodate the London Philosophical Society's proclamation (1811) that its intention was to 'remove that barrier erected by pedantry against universal knowledge which has rendered it the territory of a sect rather than the promise of the world . . . ', or David Robinson's somewhat later fear (1825) that the 'Aristocracy of Science . . . is to be the enemy and ruler of the old'.[31]

One important characteristic is shared by Thackray and Berman. The basis for Berman's generalizations was a study of the Royal Institution of London, while those of Thackray derive from his very thorough study of the Manchester Literary and Philosophical Society; therefore both owe everything to very specific and singular contexts, and both suffer accordingly. The paper on Bristol by Neve (Chapter 6 below) shows very clearly how detailed research upon other areas and institutions forces the social historian to develop a more subtle and yet embracing explanation for the popularity and vigour of science culture in these years, and particularly points to the need to recognize and accommodate contingent variations in the specific contexts of that culture.

III

The empirical and thematic pluralism incorporated in a collection of essays which range from London to Newcastle to Edinburgh, or from the Royal Society of London to the Geological and Polytechnic Society of the West Riding of Yorkshire is fairly obvious. Nevertheless, among this diversity, several themes clearly emerge, and all of them seem worthy of further research by historians of British scientific culture.

All the essays demonstrate the ultimately extra-associational nature of specific scientific activity and interest. Even where it seemingly rests upon an institutional base, as shown in Weindling's paper on the Mineralogical Society of London, no contribution presupposes that the origins, activity and dynamism of a scientific association are conditioned solely by its internal programme or agenda or by forces working upon its members from within. Although MacLeod's object is to explain the reform movement in the Royal Society during 1846–8 his subject is the period 1830–48 treated in the widest manner. Dispensing with simplistic notions of "reform" and "professionalization", he shows that the events of 1830–48 illuminate the 'process by which traditional loyalties, based upon Crown and Party, were eroded and replaced by new contractual allegiances based upon service to knowledge and utility to the State'. For good reasons the reform of the Royal Society paralleled developments in the management of English society. But while MacLeod is willing to suggest such parallels he clearly shows how party lines were blurred, and at times invisible, because of the interlock of reform with such major issues in the Society as the source and nature of patronage. Utilizing this extra-associational political schema he firmly identifies the inadequacy of such distinctions as amateurs and professionals, reformers and conservatives, when used as explanations for the reform movement in the Royal Society. In an analogous way, Orange in his study of Newcastle shows how its major science institution was some social function of the tight network of interrelations, and loyalties which the dissenting substructure manufactured. The network of Warrington Academy and Manchester New College, of the Turners, the Hollands and the Martineaus was of as much, if not more, importance in determining the history of the Literary and Philosophical Society as were specifically institutional or internal factors. The Edinburgh Philosophical Association, founded significantly enough in 1832, is depicted by Shapin as a child of petty bourgeois needs and demands which were manifested in local institutionalized competition and conflict. In his study of Bristol science between 1820 and 1860, Neve outlines the manner in which the local scientific culture developed in a mode more introverted, more conservative and generally less public than in industrialized or industrializing urban centres. An urban historical approach allows Neve to explain internal features of the Bristol Institution, for instance, the lack of "utility" of its science, the alliance of Unitarians and Anglicans in its membership, in terms of such non-institutional forces as the "conservatism" of the area and the confidence of its local social elite, in a milieu of economic decline. Such was the setting for an imitative scientific culture dependent for sustenance on its relations with Oxford and the metropolis, as encapsulated in the very name and figure of John Scandrett

Harford, FRS, landowning merchant and banker, Anglican, biographer of Hannah Moore, correspondent of Whewell and despoiler of the reputation of Thomas Paine.

A second feature common to most of the essays is that they link scientific careers to the growth of institutions. At this level, science possessed a very immediate utility, particularly for medical men and teachers. This is especially well illustrated in Hays's paper on London science lecturing, which shows that metropolitan science institutions were constantly invigorated by the existence and influx of a large medical community of both activists and audience. Emphasizing that London scientific institutions underwent challenge and change on a scale equivalent to that of the provinces, Hays argues that formalization stabilized the incomes of men of science and raised their status and seeming qualifications. When we allow that this formalization was also a feature of the provinces at this time, it becomes clear that to be a "Fellow of this" or a "Lecturer to that" served to enhance reputations and could be used to focus the attention of the public via advance publicity in areas otherwise ignorant of the peripatetic adventurer. In a somewhat different manner, this point is also broached by Durey in his assessment of the impact of the cholera epidemic of 1832 upon the self and social image of provincial medical practitioners. The fact that the episode 'highlighted the structural tensions within professional medicine' at a time when to innovate threatened exposure and failure rather than promised prestige and reward, may have been the reason why many medical men, attempting to gain a measure of social and intellectual composure, entered as activists into the scientific associations of the period. In Edinburgh, the Philosophical Association acted as a centre of activity and reward (annual receipts reached £1000) for hired and itinerant savants such as John Murray and George Combe. Indeed, the controversial plan for science lecturing to the Scottish provinces, emanating from Edinburgh in 1835, was designed to 'breed up a few young men with faculties filled to keep the fire awake', that is, as a self-sustaining centre for the diffusion of Combean science to the Scottish periphery.

A theme which has already become obvious is that of the practical utility of science as seen by both its proponents and its audience. Weindling lays stress on the very real utility of science to the members of the Mineralogical Society; corresponding members included David Mushet of the Clyde Iron Works, John Wilson of the Calder Iron Works, and White Watson the geologist and mineralogical surveyor, none of whom were mere "cabinet mineralogists". In the provincial context, the programme of the Geological and Polytechnic Society of the West Riding included the identification of Yorkshire's coal seams, their disturbances and dislocations, the compilation of statistics on coal-mining, research on the chemical structure of coal, the examination and development of techniques involved in the ventilation and draining of mines, and research into the properties of iron. Of the papers read in the period 1838–43, one third were devoted to mining technics, slightly less than one third to the geology of the West Yorkshire coalfield, and one tenth to problems of iron manufacture. Significantly, the Society was founded at a meeting of the West Yorkshire Coal Owners Association and its early utilitarianism was undoubtedly determined by the occupational interests of such leading activists as Wilson and Morton, both of whom were engaged in working deeper mines than the average. From its foundation in 1837, the Society was thus purposefully intended as a break with both the "polite" tradition and with the earlier dependency of the region on the interest of London or the more general geological activities of the philosophical societies in Leeds, Sheffield, and elsewhere. While it might be tempting to explain the decline of coal geology and other utilitarian pursuits in the Society, in terms of fluctuations in the influence of different types of leadership, it could be argued that this supposed cause was itself a result of more systematic forces. Perhaps, as Morrell suggests, an intellectual interest in science was, for the entrepreneur and risk-taker in this field, ultimately incompatible with his belief that secrecy and special knowledge were required for success. Also, the utility of geological knowledge remained attractive as a notion, but activists perceived that, as time went on, the isolated and relatively general voluntary cultural association was an inadequate body for sustaining collective research programmes.

A fourth characteristic of the essays is their emphasis on detailed empiricism which is designed to reveal the tensions, ambitions and conflicts which produce the dynamic of associational development and which operate both within and without the institutional setting. Shapin's study of the 'boundaries of participation' in Edinburgh's scientific culture reveals that conflicting groups were drawn into a scientific enterprise for divergent reasons, eventually producing a crisis when institutional change threatened the social stance of key figures. By detailed research Shapin demonstrates how cultural entrepreneurs marketed their products across barriers to diffusion, and in particular how their socially partial projects were presented as neutral cultural enterprises. A dominant tactic, which we meet elsewhere, especially at times of overt socio-political strain, was to conceal the social meaning and message of the scientific association by incorporating the allegiance (if not the effective participation) of either acceptable or counterbalancing individuals. Through such detail it is clear that the diffusion of ideas is not an asocial, passive, enervated process: 'The success or failure of diffusion hinges on contextual perceptions of science as a potential resource; its instrumental potential is assessed both by groups engaged in facilitating its diffusion and by groups on whose approval or co-operation the successful diffusion depends.' In this sense the paper illustrates the value of the approach advocated by Barnes (see Section II above). Through similarly detailed examination of the impact of ideas, Durey also shows how the presentation of knowledge is conditioned by very immediate social conflicts and interests. He focuses on a crisis situation and the associations (district and divisional boards) are crisis-bound. As a result the conflict between general practitioners and physicians in their presentation of scientific knowledge was confused and possibly reduced by the conflict between the interests of the local medical men at large and those of their local merchant and industrial clientele. The eventual common resort to traditional knowledge (remedies) and its public presentation was a counterpoint to the medical communities' accommodation to local politics: both served to reduce their collective profile during the crisis.

Leaders and activists exist within groups. A major point emerging from the collection is that such leaders both represent groups and adopt tactics whereby they may solidify and at times extend the operations and influence of particular social groups. It was leadership which sharpened the crisis in Edinburgh's petty bourgeois science during the 1830s, and it was a lack of local leadership which allowed the dilution of utilitarian science in the Geological and Polytechnic Society of the West Riding. But the best illustration of the role of the cultural entrepreneur in a particular locality is in Orange's study of science in Newcastle, which by stressing the activity of the Reverend William Turner (1761–1859) intends to 'discourage premature or facile generalisation' about urban science culture. It was from his association with the Warrington and Manchester savants that Turner derived his belief in both the utility and the radical rationalism of the scientific enterprise: 'Truth was attainable.' Such intellectual and cultural confidence allowed him to meet local opposition to science (as in 1802 and 1808–9) and to adopt tactics for its diffusion (as with his formation of an inner scientific circle which gave continuity of support). Orange tends to confirm Shapin's conception of the movement of ideas as a dynamic and at times conflict-generating process.

A sixth feature of the material in the volume is that it reveals the complex relations between cultural institutions, the development of specialization, and the encouragement of creativity and scientific mobility. Gilbert was one of those who realized that the narrow defeat of the "reformers" in 1830 would shape the pattern of things to come in the Royal Society. The victory of the Benthamite principles of 'public utility, open competition and efficient administration' derived from pressure from London's specialist scientific societies as well as from Edinburgh and the provinces, and all served to increase the potential creativity of England's principal scientific association. At the other end of our period, Weindling's purpose is to show the manner in which, during the years 1799–1806, the Mineralogical Society of London underwent transition from an informal general science association to a formal specialized institution with an established agenda for collective research. In a valuable essay he suggests the richness of the scientific "foreground", in which scientific careers emerged. With its emphasis on mineralogical surveying and collective research, the mineralogical programme of the Society was eclipsed by the schemes of the Royal Institution and others, which in turn presaged the activity of the London Institution (founded 1805) and the Geological Society (founded 1807).

Such themes by no means exhaust the exegetic value of the essays within the present volume. Doubtless other commentators would discover other common features at either the empirical or methodological levels. However, two further points are worth brief mention. The contributions coincide in suggesting that the later 1820s to early 1830s period – the era of widespread reform in British social and political institutions – was one of cultural pluralism and conflict in and between scientific associations. This is the period of specialization and challenge in the metropolis (MacLeod) and of alternatives to established philosophical associations in the provinces (Shapin, Hays, Orange, Neve). Perhaps it was this which determined the timing of the rise and fall of the Mechanics' Institutes? A second point which emerges is the importance to the scientific enterprise of what we might call (blatantly to coin a phrase), 'cultural overhead capital'. The overhead and annual costs of cultural associations (Hays, Morrell, Orange) engaged in science were significant determinants of variations and fluctuations in continuity, membership and, possibly, the scientific interests of associations. Economic considerations have been hitherto neglected by social historians of British scientific culture. The nature and timing of a scientific project could be profoundly affected by cultural overhead considerations. Thus William Turner, in projecting his plea for a New Institution for Public Lectures on Natural Philosophy at Newcastle in 1802, relied upon the timely demise of Thomas Garnett, whose apparatus was purchased for £455. As Turner put it:

To collect an Apparatus sufficient for the illustration of every branch of Physics, must be the work of time. In the mean while the loss which the literary world sustained by the untimely death of Dr. Garnett, appearing to present an opportunity of procuring the essential articles of a Philosophical Apparatus, at once, upon favourable terms, the Committee directed an enquiry to be made concerning it . . . . They have received a valuation made by professional men of first respectability . . . the commencement of the lectures will not be later than the middle of March next. . . .[32]

And the expense continued: apart from the large costs of a lecture room and laboratory, Turner himself was paid 200 guineas for his first course of lectures, and £157 for his third. It is hardly surprising that this tactical savant should so stress the industrial and commercial utility of scientific knowledge.

But if a commentary emphasized only similarities between regions and associations it would do so at the expense of the nuances present in the material. Already, in this summary and Sections I and II above, we have noted differences between contexts in terms of social class, motivations declared or otherwise, networks of individuals, and agendas of institutions. It will now be argued that such differences in the *nature* of the scientific enterprise in different contexts were related to differences in the socio-economic structure of urban settings and the consequent differences in the relations between them and the metropolis. The confines of an introductory essay force the presentation into a simplified and schematic twelve-factor analysis. It will be noted that none of the twelve factors formally expresses the role of urban space itself. It may well be true that urban space and structure mediate the perception of individuals and groups within the city, if only through the subjective construction of "mental maps"; but it is not at all clear that knowledge of the spatial characteristics of an urban area in this period will allow any prediction about its scientific culture. Through a heightening of perceptions of social differences between groups and classes, high-density living might promote conflict and the use of science as a cultural weapon, but at the same time it may be argued that high density is more likely to exist in situations where class alignments are most clearly polarized and where the growth of a scientific culture is thereby actually retarded or hampered, as may have been the case for Bradford prior to the 1830s.[33] For Edinburgh, the subject of Shapin's study of institutional conflicts, the wholesale movement of both gentry and middle class into the New Town from the 1770s to the 1800s may have further *reduced* tensions between them as they became one spatial entity.[34] During 1815–30 the social gulf between the "two" towns widened, and it is noteworthy that Shapin's study aligns the gentry and upper middle class as one group in the New Town, against the tradesmen and lower and newer professionals in the Old Town. Thus it might be maintained from a spatial perspective that the scientific enterprise in Edinburgh was conditioned, not by conflict between the upper and middle classes, but between the upper middle groups and the petty bourgeoisie. Finally, a social historian might well argue that such spatial characteristics of an urban area are themselves a result of more underlying and historical socio-economic factors.

The first two determinants of differences are demographic, relating simply to the size and rate of growth of the urban population. While it might be postulated that there existed a minimum size below which an active scientific culture could not exist continuously, above this threshold, *size* may have been of less importance than the growth rate of the population. If we accept as a rule of thumb Law's guide that the minimum size for an area to be regarded as a "town" is 2500 (meaning, for example, that Middlesbrough was "born" in the 1840s), then the total number of such areas in England and Wales rose from 256 in 1801 to 521 in 1851.[35] In 1801 the majority (62.5 per cent) were below 5000, in 1851 45 per cent were in that category and 26 per cent contained population in the range 5000–10,000. Thus Bristol with 100,000 in 1801 was enormous; and Norwich with 40,000 in the 1780s, a decline and rise to 50,000 in 1821, was one of the largest of towns. Yet both seemingly exhibit a fairly "retarded" scientific culture. They shared a stagnation in overall population growth. On the other hand, Derby was one fifth the size of Norwich in the 1780s, reached 20,000 in the 1820s and some 40,000 in the 1840s; throughout it demonstrated a very visible science culture, from the Derby Philosophical Society of 1783, through the Derby Literary and Philosophical Society of 1808 to the Derby Literary and Scientific Society of the 1840s.[36] At the simplest level, then, we might propose a difference in the scientific history of fast growing industrial areas and large but laggardly urban areas during the years from 1780. Briggs has gone so far as to suggest, admittedly as a rough proxy only, that the successive rise of provincial centres may be traced in the opening of major scientific institutions, for instance, Manchester 1781, Liverpool 1812, Sheffield 1822 and so on.[37] At the other end of the scale, the minimum population for the sustenance of a reasonably viable scientific culture is difficult to specify. Doncaster, with a population of 5000 in 1801 and twice that in 1831, certainly boasted a thriving scientific enterprise and a buoyant audience for science itinerants.[38] The peculiar situation of Warrington, with a population stagnating at 10,000–13,000 between 1801 and 1821, allowed a very rich science culture, ranging from the active Warrington Institution, to the Warrington Botanical, Natural History, Lecturing and Phrenological Societies.[39] Obviously, the heritage of the Warrington Academy and the proximity of Liverpool and, to an extent, Manchester influences offset demographic features which may have worked more potently in a relatively isolated area such as Norwich. Certainly, the lower threshold was pushed down by the cultural and personal dependency of such small centres of science activity as Chesterfield, Huddersfield and Kendal.[40]

Population features are obviously of some but not sufficient importance in explaining the cultural geography of science, and are at any rate a reflection of economic trends. Under the general theme of socio-economic features, a differentiation of science culture may be considered in terms of industrial structure, occupational characteristics, class structure and commercial openness, that is, the extent to which an urban area was subject to economic cycles emanating from market forces external to it. Given size, these four features set the conditions for most institutional developments.

The industrial structure *per se* acted more or less as an incentive to the emergence of a utilitarian science. The large-scale industry based on mineral resources – iron-smelting, chemicals, shipbuilding – was the setting for the utilitarianism of Newcastle's science as depicted by Orange. Turner stressed in 1792 that the philosopher could assist the coal viewer 'by communicating useful hints concerning the nature of the several damps and vapours which infect the mines, with the view of destroying or removing them', a theme extended to a variety of industries in his address of 1802.[41] Conversely, the polite science of Professor Charles Daubeny and Sir J. E. Smith at the Bristol Institution is explained by Neve in terms of the waning of a commercial elite. Such science was a far cry from the public lectures on metallurgy by Dr Thomas Olivers Warwick or the papers delivered by Richard Sutcliffe on the 'process for the preparation of White lead most economical and least dangerous to the operator' in the city of Sheffield, where urban growth was founded on the metalware industries.[42] In the sister town of Birmingham, a contemporary described the similarly local orientation of scientific interests in the Philosophical Society: 'The various lectures that have been delivered by the different fellows of this society, on mechanism, chemistry, mineralogy and metallurgy, have produced very beneficial effects, and contributed in a considerable degree to the improvement of gilding, plating, bronzing, vitrification and metallic combinations.'[43] A slightly later observer emphasized that 'the attendance is usually numerous and rather brilliant'.[44] It is perhaps worth noting that this society prospered despite vigorous competition from the Physiolectical Society of Birmingham, founded in 1803 'for the purpose of improving its members in natural philosophy by lecture, experiment and discussion'.[45]

But of even more importance, the industrial structure of a region determined its occupational characteristics, through the nature and scale of production, the relations of production, the extent of local economic mobility, the size of class subgroupings, and levels of skill required of workers. Birmingham, with over 500 trade categories in 1849, provided a living for a very large middling group, ranging from the shopkeepers of the petty bourgeoisie, to the artisans of the labour aristocracy and the surgeons of the lower professions. *Ceteris paribus*, scientific culture in Birmingham might be expected to have been diffused on a wide level, pluralist in its institutional composition, and used actively as a cultural signal. Occupational diversity ensured that class formation was not so polarized prior to the 1840s as to prohibit cultural competition; Briggs has compared the Birmingham and Sheffield Political Unions as representative of a 'political co-operation in radical causes' which lent richness to the political and cultural development of such regions.[46] As a result of such considerations, Briggs stresses the existence of 'separate provincial cultures'.[47] In addition, of course, this mode of industrialization provided a base of literacy and knowledge for the local scientific enterprise. To adapt from Read, the middling groups represented the non-commissioned officers of a popular science. As Neale has emphasized, within many cities there existed a greater distinction between artisans and the poor than between artisans and their employers or professional groups, and, therefore, it might be expected that local variations in the position of the artisanary and petty bourgeoisie were of prime importance in explaining cultural differentiation.[48] Shapin's contention that social class was both stratified and spatial is a result of his analysis of institutional conflicts in Edinburgh, where the cultural forms during the 1830s were ruled over by the clerk, the self-employed artisan, the printers and the jewellers. In contrast, the picture drawn by Neve of science in Bristol suggests how its weakness was the result of a lack of support from the lower middle class which took little part in the local commercial mode. Social threat was at times quite visible, with the riots of the 1790s and 1831, but the retrenchment of the elite in Bristol prohibited the cultural competition which gave rise to the diffusion of science elsewhere.

It is clear from the general literature and from the studies in the present volume, that the professional or professionalizing occupational groups were of quite disproportionate importance as cultural activists. In addition, their presence as activists was more likely in those urban centres which provided the industrial basis for a large subprofessional middling group. It is also noteworthy that Sheffield's occupational structure allowed the incorporation of diverse groupings in its dominant science association during the 1820s and 1830s. In contradistinction, Liverpool's scientific culture of that period underwent greater strain and this resulted in a more differentiated institutional framework. The Liverpool Royal Institution became a central association in the city by the 1820s and was only loosely related to the developing alternative network for scientific activity and interest.[49]

'Urban life tends to divide the proletariat from the middle classes.' Perhaps surprisingly, Engels's rule for the 1840s is followed by today's urban sociologists. In their *British Towns*, Moser and

Scott use a multivariat analysis in order to typify urban units; they discuss size, regions, population structure and social class, and argue that the most important *single* variable which differentiates towns is the last of these.[50] Several of Thackray's criteria for depicting Manchester's localism might be reduced to that of class structure, for example, the political impotence, isolation, and social fear of cultural leaders (but see Section IV below).[51] Class was no simple function of industrial and occupational structure. To use Perkin's felicitous phrase, 'class inheres in class conflict'. Conflict in our period centred on the relations between the two new industrial classes, but also arose between them and 'a ruling aristocracy which denied their right to exist as classes with their own institutions and leaders'.[52] As is shown in this volume in the essays of Shapin, Durey and Neve, institutional conflict reflected class formation.

In his typology, Foster suggests that one reaction to 'capitalist permanence' was the formation of urban subcultures through subgrouping.[53] The predominance of this phenomenon has served to confuse the concept of social class in this period, where some definitions of class attach themselves to structures, and others to secondary phenomena (see Section IV below). For present purposes, it is sufficient to note that a "subgrouping" reaction would bolster a vigorous pluralism of scientific associations, while class polarity or fragmentation would tend to reduce secondary cultural formations. The problems of the working-class science movement during the 1840s were generally a result of class formation, with the working class adopting independent urban non-science styles.[54] In a town such as Bradford, overt class antagonism at an earlier date served to delay the development of scientific associations, and this quite possibly represented a working out of the forces which Foster depicts for Oldham.[55] In those larger towns, such as Manchester, Leeds, Nottingham and Liverpool, where the middle-class activists were spatially removed, overt class antagonism may have served to ensure that the institutions of the science activists remained dominant; that the activists acted as an elite and reduced cultural pluralism.

The working out of such complex relations was most obvious during times of cyclical economic depression such as 1815–20, 1830–2, 1837–43, 1848 etc. Indeed I suggest that centres of overt class friction, which were also especially subject to economic fluctuations, would exhibit a science dominated by an urban elite; and that any tentative cultural alternatives suffered periodical setbacks which coincided with cycles of relative deprivation during which "cultural overheads" could not be maintained and the level of class conflict made culture conflict irrelevant. Depression, unemployment, crime and political resurgence coincided in Leicester and Nottingham from 1811 to 1816, in Manchester and Leeds during 1817–19 and 1842. As early as June 1837, at the beginning of the depression, there were some 50,000 unemployed or underemployed in Manchester. It is clear that in these periods science culture, especially that of the non-elite or that which was alternative to more established but non-elite associations, took on a sporadic form. Political agitation supposedly prevented public science lecturing, and unemployment or loss of real income destroyed the economic base of small scientific institutions.[56] If the industrial structure of the urban area could provide some cushion to the cyclical forces, as has been argued for both Sheffield and Birmingham, the effect may have been mitigated.[57] But even in such cases depression profoundly affected science culture. Between 1837 and 1843 Sheffield underwent its most severe industrial and trade depression: by 1843 total relief was being given to some 7000 people, and Pollard has estimated that, during 1842–3, 18 per cent of the town was unemployed and 63 per cent partially employed.[58] In such circumstances local political agitation became increasingly a reflection of solidified class interests and hitherto marginal cultural and occupational groups became firmly of the bourgeoisie. If political institutions were then more clearly defined as class platforms, so too were such associations as the Mechanics' Institute and the Philosophical Society. During the 1840s the previous science pluralism spawned an increasingly introverted elite science, the beginnings of a specialist civic science, and a prolific non-science working-class culture which existed alongside an explosion of vastly diluted science as entertainment. Science mutated rather than failed.[59]

We may also isolate a second group of far more diverse forces which aided cultural differentiation: socio-economic factors operated within an environment which included relative geographical isolation, earlier scientific traditions, non-scientific (for example, religious) cultural groups and institutions and the presence of "cosmopolitan" groups and individuals. Given the existence of classes and groups, it might be maintained that the variety of religious, political and other cultural features (within classes and regions and between groups) would be to an extent a function of spatial and occupational factors. Nevertheless, in areas like Sheffield or Newcastle, science benefited enormously from the fact that it was representative of and publicized by highly distinctive socio-cultural groups. Furthermore, if it were argued that science was especially attractive for, let us say, particular religious sects, then we would expect systematic variations between urban areas in either the magnitude of science or the motivation for it. Where cultural distinctiveness was a heritage of earlier eighteenth-century development, then interesting anomolies arise with respect to demographic factors: despite stagnation in terms of population, York exhibited a thriving scientific culture throughout this period.[60] Finally, areas such as Edinburgh, Manchester and Liverpool, which were centres of regional subsystems, produced "cosmopolitan" types whose cultural perspective was as much outer as inner directed. The ambivalence of the socially aspiring in the provinces was increased wherever their focus was metropolitan. Merton's 'disjunction between culturally prescribed aspirations and socially structured avenues' for the realization of aspirations was more significant when geographical restrictions were also operating.[61]

The eleventh element operating upon urban science was the most random of all. The persistence and strategic success of leading savants determined much that otherwise would be difficult to explain, especially in the less established urban cultures. Bristol had no William Turner, but Hull had its John Alderson, Derby its Erasmus Darwin and Warrington its James Kendrick.[62] Savants like Turner or Alderson not only promoted an existing interest in science but also answered the arguments of its critics and stimulated the apathetic. They advocated science by appealing to the spirit of trade, the spirit of progress and the spirit of the region itself, and were by no means necessarily allied with the local elite. As Alderson expressed it, in an address delivered from his newly-established science lecturing institution at Hull in 1804:

... from this establishment more learning has been diffused throughout the Town, than could ever have found its way along the private channels of a few rich individuals. Into this room, as into a focus, I would endeavour to collect every ray of intellectual light; and from this place, as from a centre, I would have it emanate to every point of the neighbourhood.[63]

A last and important feature of urban structure which affected the shape of the scientific enterprise was that of local political structure. Although we may write generally of the rise of class and of reform, the actual direction taken by urban pressure groups was very much conditioned by the local political constraints within which they could operate and within which they formulated their wider political convictions and programmes. Local politics in turn affected the position of cultural leaders and their attitudes to natural science, together with the propensity of their audiences to grow or diminish. In unreformed boroughs, local electioneering and agitation produced antagonism between the middling groups and the local gentry or established merchants, and thus spawned subgroupings. According to Fraser, it was during these years that minor institutions (for instance, the vestry) were 'politicised' as new urban groups sought out every layer of potential political influence.[64] Perhaps it is not merely a coincidence that the period of particular pluralism and energy in provincial science occurred alongside the agitation for local reform in the 1820s and 1830s. When reform did come to the older conservative boroughs of Bristol, Hull, Liverpool, York, Norwich and so on, it had a proportionately greater effect in redefining the position of key social groups. Ironically, the civic science of the 1840s was one result. As early as 1835 when standing for a ward under the new Act, James Carson of Liverpool, a leading reformer in the town, president of the Medical Society, active in the Natural History Society and, at that time, engaged in a series of lectures on acoustics and optics before members of the Mechanics' Institute, had clearly set out the connection between civic science and reform in municipal government:

All questions relating more particularly to the Public Health, such as the removal of nuisances, and the prevention of deposits or works which may vitiate the atmosphere in a way that may be detrimental to life . . . and other sanitary objects, as my mind may be supposed to be *professionally* habituated to them, will naturally be expected to engage my peculiar care. . . .[65]

The success of the shopocracy in the council elections of 1838 in Birmingham and Manchester similarly paved the way for a new alignment of social forces and a new place and image for the natural sciences.[66]

The variation of spatial and socio-economic forces between provincial centres meant that the provincial–metropolitan relationship, while it existed and had significance, was not the same for all regions. It is true enough that it was the existence of London that made the provinces "provincial", but Briggs is nevertheless correct when he writes of 'separate provincial cultures'. It would be convenient to depict the metropolitan–provincial relation in terms of, say, resentment prior to the 1790s, emulation from the 1790s to the 1800s, a return to resentment from the 1800s to the 1820s, to be rounded-up with a prideful independent provincialism during the 1830s and beyond, as epitomized in Cobden, Bright, O'Connor and the Anti-Corn Law League. In such a schema Read's recognition of provincial non-conformists' 'readiness to organise'[67] might be seen as a cause of their involvement in science during the "resentment" periods, when the national political economy did not yet allow them expression through formal political activity, followed by a withdrawal from science during the "provincial" phase. But we have already shown that such an approach is unworkable. Nonconformists did not hold a monopoly over science. Urban science culture involved an audience and was not therefore solely determined by the needs and activities of "cosmopolitan" intellectual leaders. The civic science of the post-1830s was provincial in tone and replaced the perhaps more radical culture of the earlier years. Even if the period of "emulation" existed, it by no means excluded a rise of provincial science, if only because London science was also undergoing change and reorganization (Hays, Weindling).

Variations in the provincial–metropolitan relationship appear both extreme and endless. Neve shows how the elite which dominated Bristol's science was by no means "cosmopolitan" in Thackray's sense, but was nevertheless closely related to both Oxbridge and London, and tended towards imitation (that is, emulation) of metropolitan forms and conventions. As he puts it: 'Both Bristol and Bath may well have been trapped in a dependence on metropolitan forms which only industrialisation could break, and industrialisation left few marks on the economy of the West Country.'[68] In industrial and isolated Newcastle, the distinction between practical and polite science was equated to that between province and metropolis. Here was a virile provincialism which reflected resentment and even competition more than emulation. Turner was not above invoking international competition as an incitement to action. Admitting the strength of metropolitan science societies, he:

regretted, that, while, in Germany, France and Italy, there is scarcely a provincial town of consequence which has not some establishment of this kind, in England they have been, in a great measure, confined to the metropolis. . . . It is obvious that Newcastle enjoys peculiar advantages for chemical investigations. . . . It will be a worthy object of such a society to enquire, how far the country is still *improvable*. . . .[69]

The model subsequently adopted at Newcastle owed more to emulation of Hull and Manchester, and to the provincial science savants such as the public lecturers Moyes, Stancliffe and Garnett, than it did to developments in metropolitan institutions.[70] Porter suggests that the geological and mining interests of the Newcastle scientific community were an obvious result of industrial interests and a felt resentment against the metropolitan dominance over geology, which became solidified in the division of the membership of the Geological Society into metropolitan savants and provincial "collectors".[71] In his essay on Yorkshire geology, Morrell emphasizes how, in the 1830s, the 'metropolitan coterie' was socially distant from the surveyors and coal viewers of the provinces. In such instances it might be maintained that tension between the centre and the periphery in scientific organization lent extra vigour to the latter.

Edinburgh – province or metropolis? In his treatment of scientific culture, Shapin answers the question by labelling Edinburgh 'a provincial metropolis', that is, the metropolis of Northern Britain, and thereby a cauldron for the manufacture and export of science as a product. As with London, Edinburgh created its own science on a very significant scale and then exported it to the cultural "wastelands", as the Scottish metropolis saw the provinces until at least the mid 1830s. In this situation, the clash between groups within Edinburgh at times became a clash between metropolitans and locals, and this was probably more clearly expressed than was the similar difference of perspective between "cosmopolitans" and "locals" in such English urban centres as Manchester. One reason why the Edinburgh Philosophical Association opposed Combe's proposal for diffusion of science in the hinterland was that it focused attention away from the local and immediate needs of Edinburgh's petty bourgeoisie. Savants in Edinburgh were forced into "tactical alliances" which were intended to reconcile their essentially "national" objects and pursuits with the needs of local interest groups and audiences. 'Edinburgh. . .was a substitute London, providing metropolitan amenities in a Scottish context.'[72] The rise of Edinburgh dominated Scotland. When noting the general weakness of Scottish provincial science, Porter specifies the effect of this on Scottish geology. Because geology in Scotland was directed by a 'closed community of Edinburgh naturalists, gearing their science to university teaching needs...', it remained less practical, empirical, provincial and vigorous, and more highly 'philosophical' than its counterpart in England.[73] First and foremost, talent found its way to Edinburgh.

Simultaneously, there is some sense in defining Edinburgh as a provincial centre. Smout has argued that in Scotland 'the nobles and gentry were as cosmopolitan as they could afford to be': their focus on London rather than Edinburgh prevented insularity.[74] Given that professionals of all sorts dominated Edinburgh's intellectual life – lawyers, teachers and medical men, in that order of importance – and given that they sought patronage from the gentry, as well as felt a "deep-seated need" for their approval, then it is hardly surprising that Edinburgh, its university and its cultural activists, were profoundly cosmopolitan at heart. If provincials saw Edinburgh as an attraction, then those who succeeded in Edinburgh focused on London, and through such channels Scotland 'became more British'.[75] Improvement, through associations or otherwise, opened the road to the south. After 1830, coincident with the relative decline of the university, Edinburgh failed to contain its talent. Scott may have died fighting, but Carlyle took the London coach. If past investment in cultural overhead capital was impressive 'the result was an intellectual vacuum in the north'.[76]

London was thus affected by both supply and demand forces emanating from the provinces and Edinburgh. Although London's scientific institutions absorbed provincial talent, they were also forced to recognize and respond to the provincial challenge. Yet on many fundamental levels this was the period of London's retardation. For the first and last time in history London lost its demographic domination over the nation.[77] Industrially, London's industries failed to compete with those of the provinces, and were only partially compensated for by the rise of newer and lighter consumer industries. More significantly, vested interests held back municipal reform (for example, the implementation of the Act of 1835 or the Public Health Act of 1848) and promoted a framework for the fragmentation and disintegration of metropolitan public opinion. The lack of peculiarly local grievances in London meant that the 1790–1830 period was one of provincial political leadership.[78] Referring to the crucial years 1830–2, Shephard has stressed that 'when compared with Birmingham, a town with less than one tenth of the population of the metropolis, the impact of London on the course of events seems to have been relatively small. . . '.[79] One may suggest that, alongside the rise of administration, trade and merchandising as responses to economic threats, there was a rise of cultural authority in response to provincial political and cultural threats. But where in industry and politics, by their very nature the constraints on response were strong, in matters of science culture, the metropolis underwent a rejuvenation. In addition to that originating in alternative and specialist metropolitan societies, MacLeod discerns some of the early "attack" on the organization of the Royal Society as being located in the scientific communities of the provinces, confident of their 'important new traditions'. MacLeod quite clearly centres on the 1830s as the provincial decade in both science and politics, linked to reform in the Society through the British Association for the Advancement of Science, the "social laboratory" which acted as a "commons" to the Society's "upper house". Through the important transitional decades of the 1820s and 1830s, by the 1840s the enthusiasm and public interest of the provincial societies had generated formidable rivals.

In Scotland, the dominance of Edinburgh and its ascendance as a metropolis retarded provincial cultural development. In England, the independence and competitiveness of the provinces was in part a function of their *provincialism* and London responded to ensure its continued leadership of the scientific enterprise. The gulf between the Scottish periphery and its centre was dysfunctional, while in England it appears to have been optimal for the development and diffusion of a vigorous science culture. Moreover, in its second role as a British province, Edinburgh acted as an entrepôt for the talent which found its way to London. In Edinburgh, value was added to the raw material from the provinces and the semi-finished product was removed through neo-colonial channels to the true centre of British intellectual life. Thus, English cultural imperialism acted as a final arbiter of the difference between the scientific histories of the two nations, especially in so far as those histories were conditioned by metropolitan–provincial cultural relations.

IV

Most of the writings surveyed in Sections I, II and III have implicit within them some notion of the social system underlying British culture at this time, its nature, and its functional relation to that culture, and particularly, the role of social class in interlinking the two. Such relations may lead to confusion, as with Thackray's comment that his provincial savants, 'were among the larger, but uncounted, group of English provincials who successfully *sought* the heightened social standing and envied cosmopolitan connections implicit in the Fellowship [of the Royal Society]', a phrase which sits uneasily with the same writer's conception of natural science functioning so as to underwrite 'their distance from the traditional value system of English society . . . '.[80] Was there indeed a mature class structure to which science culture, as it were, functionally conformed? If there were two "methodisms", were there also two "sciences"? If the answers are affirmative, does it follow that the aim of members of the wider scientific movement was to use their allegiance to science, among their other social actions and attributes, as a means of upward social mobility? Was the latter perceived as a process of social internalization, towards a social centre or apex which had significance as either a legitimater or awarder of social status? It can be argued that, because the answers to such questions are not in fact wholly affirmative, and because historical research suggests a *making* of social class and a disorganization of the "social fabric" in the period prior to the 1840s, then the motives of individuals within the scientific movement and culture, and therefore the explanation of that culture, may not be sweepingly attributed to their needs for upward mobility or for social legitimation.[81] Even if upward mobility were to be considered a prime characteristic of such groups, and this would at present seem to be the historical consensus,[82] this does not in itself uncover motivation nor does it show that the major perceptions of individuals included the idea or the image of *a* social centre towards which they could safely choose to migrate. Finally, it should be made clear that *even if* mobility was the intention and a centre *was* perceptible, then it has already been strongly suggested (see Section II above) that natural science was not the most obvious cultural form for individuals to adopt.

However, what of science as a counter-culture? If science was at sometime representative of an *alternative* social system, at what point did it cease to be so? The answer to this question depends upon the historian's assumptions or thesis about the class nature of British society during the industrial revolution. While local variations in the scientific enterprise may have resulted from the types of factors outlined in Sections II and III above, the more universal effects of urbanity and class formation may well have served to superimpose a degree of generality on trends and characteristics in the scientific culture prior to 1850 for example, its politicization in the 1790s and 1820s, the emergence of Mechanics' Institutes or their equivalents during the late and mid 1820s, the rise of a more "civic" science and its relative social introversion in the 1840s and so on. However, any remarks at this level are highly tentative: they require confirmation or refutation from further informed research of the type exemplified in the present volume.

In attempting to "explain" both the rise and the role of a popular science culture in Britain, the following explicitly class-based thesis may be offered. In Britain, the last decade of the eighteenth century and the first three decades of the nineteenth century saw the making of social class, a process which centred upon the economic and social evolution of the industrial provinces. Prior to the 1840s, when class was thus "in the making", significant social groups and institutions emerged which, in their structure and functioning, represented no mature class formation, but rather reflected the social upheaval and reorganization which was itself a result of widespread economic change.[83] In short, Britain, especially industrial provincial England, underwent a social revolution. This was the setting for the development of social groups who were essentially "marginal" to society because neither overtly of the capitalists and often decidedly not of the working masses. While some historians have seen such groups as subsets of existing classes or as the substance of an entirely separate "middling" class, the degree of social flux encourages a consideration of them as essentially marginal, amid a changing system wherein there was no longer *one* social world.[84] Park's emphasis upon the individual 'moving in more than one social world' is of especial importance during a period of significant social change and dislocation prior to the emergence of decisive class boundaries and ideologies.[85] When social change is rapid, the individual's perception of central or commanding values, beliefs and norms is likely to be both blurred and confused. During this "mass milling process", when social trends are in flux and their ultimate direction in real doubt, the establishment of a feeling of community and a *characteristic* social location or profile becomes appropriate as an alternative to attempts at social centralization or integration. Referring to the genesis of urbanization, Wirth has written of the individual that, 'reduced to a stage of virtual impotence as an individual, the urbanite is bound to exert himself by joining with others of *similar* interests into *organised groups* to obtain his ends'.[86] In effect, the problem of identity, which we have assumed to be a social product, may be resolved through actions which serve to identify the individual within a group or distinguish him as a social type (Thackray's 'distancing'), rather than actions which function directly so as to secure greater social integration. This is the case simply because the latter strategy depends for its formulation on at least a minimum degree of consensus as to the *nature* of the dominant values, beliefs or patterns of behaviour, and this in turn requires a minimum degree of institutional continuity.

Such relations have important implications for the historian's treatment of the motivation of metropolitan and provincial savants. Although the actual forms of response varied according to the extent of social change or peculiarities within particular locales, the continuous and underlying need or motive was the securing of a social identity and a feeling of belonging. The institutions and groupings of science culture were utilized by the marginal man in first gaining and then propounding his social identity. This was especially the case with the younger individuals who were active in the particularly "popular" phases of science culture – the 1790s and 1820s – during which science was far from being seen as central to the underlying values of the social system or particularly part of the ideology of a ruling class.

It is at this point in the argument that the interpretation of data is most debatable: hence Thackray and Berman, using similar material, arrive at different conclusions. For it is indeed the case that one result of the confused position of marginal men as they are here conceived might be social mobility and a movement towards an increasingly discernible "social centre", perhaps a major facet of the social history of the 1840s. But there is no need to *explain* any such trend as the result of an initial desire for upward mobility and integration.[87] For example, in joining a scientific association in 1820 the *motive* of a marginal man was that of securing a greater degree of social identity and comfort; at the level of the functionalist's conception of "system integration", the social *function* of the association was the integration of such individuals; the result of this over a longer period could be the upward mobility and centralization of initially marginal individuals. In this simple schema there will be periods when the marginal man, through his characteristic stance, is at social odds with other, perhaps more central groups in the society. A subsequent period might nevertheless find him or his successors at or near the centre of a changed social system.[88]

During the 1840s scientific culture lost such social functions. By these years, science, as an intellectual system, was radically different from the science of the 1790s or the 1820s. To borrow from Bernal, this was the beginning of the period in which the early revolutionary advances were to be 'digested'. The resultant greater sophistication and specialization had two major effects on the scientific movement and science culture. First, such trends ensured that natural science could no longer be encompassed within the format of a public lecture course, however competent the lecturer. The surviving serious activity became expert, specialist and applied (as with Kargon's 'civic' science) and the educational function of the associations was severely reduced.[89] The contemporary "polite" science of the savants and the highly popular science of and for the working class were not adequate alternatives to the earlier, socially characteristic scientific culture. Second, such changes meant that science was no longer as capable of projecting a social image, especially a radical, questioning, alternative image for those caught in an urban identity crisis. As this occurred so too was the need for an "alternative" science culture reduced. Not only were individuals undergoing social relocation by means of their own activity, at the same time many of their social attributes were being systematically redefined. The transfer of power associated with the first Reform Bill or the new religious and civic toleration represented by the legislation of 1828 and 1844 reduced the innovating, agitating role of many urban savants.[90] This may be pushed to an extreme by arguing that *the* final hardening of class lines occurred as such "buffer" groups became socially introverted or, as in the case of those who remained socially active, at one with the industrial bourgeoisie. Finally, although the new conditions might have meant that natural science could *now* become a weapon of hegemony, it is also true that no vigorous scientific culture developed around this aim, if only because the function of social control was by this ti e deing fulfilled by a vast range of cultural, educative, political and other institutions.[91] Science as such may have been too obvious as a means of control over the working class, who repeatedly rejected it within their own institutions, which gained their independence during precisely these years.[92] Perhaps it was because of such reasons that between the 1840s and the 1870s British science appears not to have produced a popular movement or image or to have been purposively cultivated by *any* characteristic social groupings to a degree even comparable with that of the earlier period. Although the relative isolation of the scientific enterprise in these later years has yet to be properly documented and analysed, it may be that such a conjuncture of socio-intellectual forces will be found to be fundamental. If Bernal focuses on the practitioners when he writes that 'specialisation itself was a way of escaping the too heavy burdens of a general view of the universe', Yaran Ezrahi encompasses the wide culture when he claims that 'it is in the incipient stage that science is most likely to appeal to the layman'.[93] But the general neglect of these interim years has without doubt led to both a false belief in the ascientific nature of the industrial revolution and to the thesis that Britain's lag in formalized scientific and technical training after 1870 was simply a continuation of a 100-year trend.

When Margaret Gowing lists five 'possible causes' of the British science-technology lag in the late nineteenth century, none exposes the social and economic effects of those profound changes in later nineteenth century industrialization which revolutionized both Britain's position in the world economy and the structure of her internal economy.[94] When Oliver MacDonach "explains" Germany's economic concentration and comparatively sophisticated "national organization" – and thus her growing supremacy in organized science and technology – on the grounds that it 'was state policy that this should be so', he neglects to add that such policy was strongly conditioned by Germany's relative backwardness in industrial development in the earlier period i.e. *c*. 1840–70s.[95] As a final instance, Harold Sharlin appears to argue that the major reason for England's backwardness in formal, government sponsored science training was her early nineteenth-century heritage of scientific individualism and institutional *laissez-faire*. Thereby he does not acknowledge that such characteristics were – and are – lauded as among the principal reasons for England's entrepreneurial zeal, freedom of economic decision-making and, so, successful early industrialization.[96]

Economic historians have moved away from the notion of the

1870s as the beginning of a "climacteric" which measured Britain's economic downfall. The British economy of the last third of the nineteenth century is increasingly seen as one wherein industrial production and productivity are declining, where industrial profits are, as a result, falling, but where unemployment is not rising, where entrepreneurs are moving into new areas (e.g. retailing) and where the general focus of the economy is away from industry and towards services, capital exports and imperialist or neo-imperialist trade.[97] Decline has become reorientation. Internationally, this maturing of the economy was juxtapositioned with the industrialization of other nations, notably Germany, the USA, Japan and Russia. As Landes has reminded us, Germany was an economy that was well behind Britain in 1870 in assimilating the technology of the industrial revolution.[98] Nor was this juxtaposition, definable as a technological lag, unrelated to the motives of key elites within the new industrializers. Hobsbawm has emphasized 'that a nationalist desire to catch up with the British was largely responsible for the systematic German reinforcement of industry by scientific research'.[99] But this is no isolated case and may be viewed within a yet wider framework of analysis. Gerschenkron has shown quite adequately how relative backwardness determines a foremost role for the state in instigating and directing the institutionalization of the development process during periods of industrial spurt.[100] Industrial revolutions were not simultaneous; and the later nineteenth century saw the interface, through flows of goods, capital, technology and personnel, of one huge industrial economy with several industrializing nations. The changing nature of best-technique technologies in engineering, chemicals, electricity, metallurgy and electro-metallurgy and their growing dependence on specific scientific knowledge, together with the availability of foreign capital and supply outlets, meant that industrialization in such later developers was faster, more reliant on sophisticated technology and the capital goods industries, and more institutionalized (e.g. tariffs, taxes, banks, land reforms and science education) than was Britain's industrial growth at *that* time or during her earlier industrialization.[101]

The question then arises as to why Britain did not adapt to the strains engendered by this juxtapositioning by increasing her industrial productivity, investing in science and technology, and rejuvenating her social and economic institutions. The brief answer is that these were not the only possible forms of action. The cries for science may have been real enough when they emanated from industrialists whose profits were falling, but that is certainly not to say that increased investment in science or technology was an optimum or even reasonable economic choice for the nation as a whole.

If the main thrust of the British economy after 1870 was a dynamic reorientation towards trade, capital exports, services, imperialism and so on, then the whole thesis of science-technology "failure" is called into question or, at the very least, requires careful restatement. Although 'scientific training developed on the continent ahead of industry' because of the nature of industrialization in conditions of relative backwardness, in Britain technology and science were *decreasingly* important as an effective constraint on the progress of the economy.[102] Furthermore, such an argument suggests that the ready growth of science-based industries in nations like Germany was *not* simply a consequence of efficient technological training or institutions, but was more a result of such factors as lower scrapping costs, a large technical lag initially which induced large increases in productivity, a changing technological imperative, and an elite-engendered industrialization drive.

V

If full of suggestions, this essay is light on confirmed conclusions. But perhaps this is as it should be. Hobsbawm quite rightly omitted the social history of not only science but also intellectual processes *per se* from the discernable 'types or approaches' in recent British social history.[103] It is equally true that among those which he did single out as vigorous and significant were urban studies, the histories of class, social groups and 'mentalities', which closely interface with the materials and methods of many social historians of British science. But the subject area is not yet defined: should phrenology be analysed in a manner similar to physics? Nor, for the period considered here, have major theses been substantiated. The essays in the present volume follow Crosland's recent advice that we 'must study institutions' if we are to discern the essence of a nation's intellectual life, and depict the 'traditions of local initiative and independence' which characterize the British model.[104] Institutions, groups and regions must ultimately be viewed (if only to be selected in the first instance) in terms of more or less clearly stated hypotheses or theories, and these have yet to achieve recognition and specificity. In the absence of such theories, this essay has attempted to show common themes and possibilities of interpretation which stem from a consideration of work so far produced, particularly the essays collected together in this volume.

Notes and references

1   T. S. Ashton, *The Industrial Revolution 1760–1830*, London, 1948, p. 16.
2   P. Mathias, 'Who unbound Prometheus?: science and technical change 1600–1800', *Yorkshire Bulletin of Economic and Social Research*, 1969, **21**, 3–16; A. L. R. Hall, 'What did the industrial revolution in Britain owe to science?', in N. McKendrick (ed.), *Historical Perspectives, Studies in English Thought and Society*, London, 1974, pp. 129–151; A. Thackray, 'Science and technology in the industrial revolution', *History of Science*, 1970, **9**, 76–89.
3   See particularly A. E. Musson and E. Robinson, *Science and Technology in the Industrial Revolution*, Manchester, 1969; A. and N. L. Clow, *The Chemical Revolution*, London, 1952.
4   Contrast R. A. Smith, *A Centenary of Science in Manchester*, London, 1883 and J. H. Plumb, *England in the Eighteenth Century*, Harmondsworth, 1950, p. 167 with A. Thackray, 'Natural knowledge in cultural context: the Manchester model', *American Historical Review*, 1974, **79**, 672–709.
5   J. W. Hudson, *The History of Adult Education*, London, 1851; T. Kelly, *A History of Adult Education in Great Britain*, Liverpool, 1962. But see also T. Kelly, *George Birkbeck*, Liverpool, 1957.
6   S. Shapin and B. Barnes, 'Science, nature and control: interpreting Mechanics' Institutes', *Social Studies of Science*, 1977, **7**, 31–74; I. Inkster, 'The social context of an educational movement: a revisionist approach to the English Mechanics' Institutes 1820–50'; *Oxford Review of Education*, 1976, **2**, 277–307; C. M. Turner, 'Sociological approaches to the history of education', *British Journal of Educational Studies*, 1969, **17**, 146–65.
7   A. E. Musson, editorial introduction, *Science, Technology and Economic Growth in the Eighteenth Century*, London, 1972, p. 62.
8   As shown clearly throughout Musson and Robinson, op. cit. (3). An acknowledgement of the known data might lead to the view that industrial and urban utility was the most fundamental and real attraction of science for not only the audience but the mass of activists also.

9   Norwich science tended to be confined in such associations as the Speculative Club and in the science provisions of the Norwich Library. For Nottingham see R. M. Bowley, F. M. Wilkins-Jones, and D. Phillips, *et al.*, *George Green, Miller of Sneinton*, Nottingham, 1976; S. D. Chapman, 'Scientists and innovators', in K. C. Edwards (ed.), *Nottingham and its Region*, Nottingham, 1966; I. Inkster, 'Scientific culture and education in Nottingham, 1800–1843', *Thoroton Society Transactions*, 1978, **82**, 45–50.

10   B. Barnes, *Scientific Knowledge and Sociological Theory*, London, 1974, p. 116. For a detailed case study in this approach see S. Shapin, 'Phrenological knowledge and the social structure of early nineteenth century Edinburgh', *Annals of Science*, 1975, **32**, 219–43.

11   M. Berman, 'The early years of the Royal Institution, 1799–1810; a re-evaluation', *Science Studies*, 1972, **2**, 205–40 (238); *Social Change and Scientific Organization: The Royal Institution 1799–1844*, London, 1978.

12   Berman, ibid.; H. B. Jones, *The Royal Institution, its Founders and its First Professors*, London, 1871; G. N. Cantor, 'Thomas Young's lectures at the Royal Institution', *Notes and Records of the Royal Society*, 1970, **25**, 87–111; H. Hartley, *Humphry Davy*, London, 1966. Accum and Garnett, particularly, owed much to a network of London institutions, and the lives of both would repay further study.

13   Thackray, op. cit. (4), 675–7, 698–707; F. Nicolson, 'The Literary and Philosophical Society 1781–1851', *Manchester Memoirs*, 1924, **68**, 97–148.

14   I. Inkster, 'Science and society in the metropolis: a preliminary examination of the social and institutional context of the Askesian Society of London, 1796–1807', *Annals of Science*, 1977, **34**, 1–32.

15   L. Williams, 'Michael Faraday's education in science', *Isis*, 1960, **51**, 515–30.

16   J. B. Morrell, 'London institutions and Lyell's career: 1820–41', *British Journal for the History of Science*, 1976, **9**, 132–46 (132).

17   By this is meant that science was "popularized" as a result of purposeful social action, and thus its "popularity" has a discernible historical dimension. See Z. Barbu, 'Popular culture: a sociological approach', in C. W. E. Bigsby (ed.), *Approaches to Popular Culture*, London, 1976, esp. p. 42.

18   A. Thackray, 'The industrial revolution and the image of science', in A. Thackray and E. Mendelsohn (eds.), *Science and Values*, New York, 1974, pp. 2–18.

19 Priestley retained his image as a 'radical republican bogy man': *Philosophical Magazine*, 1833, **2**, 382–402.

20 J. B. Morrell, 'Individualism and the structure of British science in 1830', *Historical Studies in the Physical Sciences*, 1971, **3**, 183–204 (192). Note that this stark and simplistic extension of the "individualism" approach was not that proposed by Morrell, whose paper centres on the organizational and career implications of individualism and voluntarism.

21 M. Berman, 'Essay review', *Journal of Social History*, 1972, **1**, 521–7 (526).

22 M. Berman, '"Hegemony" and the amateur tradition in British science', *Journal of Social History*, 1975, **8**, 30–50, (32, 34); A. Davidson, *Antonio Gramsci: The Man, His Ideas*, Sydney, 1968; T. R. Bates, 'Gramsci and the theory of hegemony', *Journal of the History of Ideas*, 1975, **36**, 351–66.

23 Berman, ibid., 36–7.

24 Berman, op. cit. (21), 526.

25 Thackray, op. cit. (4), 678. For encapsulation and further generalization see the same author's 'Scientific networks in the age of the American Revolution', *Nature*, 1976, **262**, 20–4.

26 Thackray, op. cit. (4), 682.

27 P. Laslett, *Family Life and Illicit Love in Earlier Generations*, Cambridge, 1977, p. 6.

28 Thackray, op. cit. (4), 697.

29 E. P. Thompson, *The Making of the English Working Class*, London, 1963, pp. 26–7 and ch. 2 generally; C. Garrett, 'Joseph Priestley, the millenium and the French Revolution', *Journal of the History of Ideas*, 1966, **16**, 50–66; R. E. Schofield, *The Lunar Society of Birmingham*, Oxford, 1963, pp. 328–69; A. Temple Patterson, *Radical Leicester*, Leicester, 1954, pp. 63–78; N. Hans, *New Trends in Education in the Eighteenth Century*, London, 1951, esp. ch. 8; E. Halévy, *England in 1815*, London, 1960 edn, pp. 577–87.

30 I. Inkster, 'London science and the Seditious Meetings Act of 1817', *British Journal for the History of Science*, 1979, **12**, 192–6.

31 *Philosophical Magazine*, 1812, **39**, 142–50; *Blackwood's Edinburgh Magazine*, 1825, **18**, 350.

32 W. Turner, *General Introductory Discourse on the Advantages of a New Institution for Public Lectures on Natural Philosophy*, Newcastle, 1802, pp. 17–18.

33 C. A. Federer, 'Ms of the Bradford Mechanics' Institute' [Local History Department, Bradford Central Library, bound MS, vol. B

874.9. Fed.]; J. Farrar, *Autobiography of Joseph Farrar*, Bradford, 1889.

34   T. C. Smout, *A History of the Scottish People 1560–1830*, London, 1969, pp. 338–9, 347–9.

35   C. M. Law, 'The growth of urban population in England and Wales, 1801–1911', *Transactions of the Institute of British Geography*, 1967, **41**, 125–43.

36   I. Inkster, 'Studies in the social history of science in England during the industrial revolution', unpublished Ph.D thesis, University of Sheffield, 1977; B. D. Hays, 'Politics in Norfolk 1750–1832', unpublished Ph.D thesis, University of Cambridge, 1957.

37   A. Briggs, *Victorian Cities*, London, 1963, pp. 46–8.

38   C. W. Hatfield, *Historical Notices of Doncaster*, 3 vols., Doncaster, 1866–70, i, pp. 473–7 and ii, pp. 330–51; *White's Directory of the West Riding*, Doncaster, 1837; *Doncaster, Nottingham and Lincoln Gazette*, 5 July 1805, 12 July 1805, 7 February 1806, 24 July 1807, 30 October 1807, 6 November 1807, 3 September 1824, 17 September 1824, 9 January 1835, 30 January 1835, 6 March 1835, 13 March 1835, 27 November 1835.

39   'Warrington Institution minute book 1811–13' [Warrington Central Library, MS 16]; H. Boscow, *Warrington, A Heritage*, Warrington, 1947, pp. 65–92; *Slater's Directory of Liverpool and its Environs*, Manchester, 1844, pp. 96–100; *Holden's Triennial Directory*, Manchester, 1809, 1811; J. Kendrick, *Profiles of Warrington Worthies*, London, 1854; W. Beaumont, 'Warrington celebreties; additions to Dr. Kenrick's Warrington Worthies' [MS bound volume, Warrington Central Library, MS48].

40   *Derby Mercury*, 5 November 1828; *Sheffield Independent*, 1 August 1829; *Sheffield Mercury*, 27 September 1834; *Liverpool Kaleidoscope*, 1828–9, **9**, 2; *Kendal Chronicle*, 21 June 1828.

41   W. Turner, *Speculations on a Literary Society*, Newcastle, 1792, p. 5; Briggs, op. cit. (37), pp. 3–9.

42   M. Brook, 'Dr. Warwick's Chemistry lectures and the scientific audience in Sheffield', *Annals of Science*, 1955, **11**, 224–37; 'Ms. minute book of the Sheffield Society for the Promotion of Useful Knowledge' [Sheffield Public Libraries, SLA 820.6.5, SLPS 216], 4 November 1804.

43   C. Pye, *Modern Birmingham*, Birmingham, 1818, pp. 37–8.

44   J. Drake, *The Picture of Birmingham*, Birmingham, 1825, p. 36; J. A. Langford, *A Century of Birmingham Life, 1741–1841*, Birmingham, 1868, **2**, pp. 368–73.

45   *Philosophical Magazine*, 1803, **13**, 86.

46   Briggs, op. cit. (37), ch. 5.

47   ibid., p. 43.

48   R. S. Neale, 'Class and class consciousness in early nineteenth century England: three classes or five?', *Victorian Studies*, 1968–9, **12**, 5–32.

49   H. A. Ormerod, *The Liverpool Royal Institution*, Liverpool, 1953; H. A. Whitting, *Aldred Booth, Some Memoirs, Letters and Family Records*, Liverpool, 1917; Inkster, op. cit. (36), chs. 5 and 6.

50   C. A. Moser and W. Scott, *British Towns*, Edingburgh and London, 1961, ch. 6; cf. P. Mann, *An Approach to Urban Sociology*, London, 1965, pp. 105–6.

51   Thackray, op. cit. (4), pp. 679–81.

52   H. Perkin, *The Age of the Railway*, London, 1970, pp. 159, 168.

53   J. Foster, 'Nineteenth century towns: a class dimension', in H. J. Dyos (ed.), *The Study of Urban History*, London, 1968, pp. 281–99; *Class Struggle and the Industrial Revolution*, London, 1974.

54   Inkster, op. cit. (6).

55   Foster, op. cit. (53); T. Steadman, *Memoir of the Rev. William Steadman*, London, 1838, pp. 363–8; *First Annual Report of the Bradford Mechanics' Institute*, Bradford, 1833.

56   H. Perkin, *The Origins of Modern English Society 1780–1830*, London, 1969, pp. 164–73; D. Read, *The English Provinces c. 1760–1960*, London, 1964, pp. 61, 110; Temple Patterson, op. cit. (29).

57   Read, ibid., p. 35; and generally G. I. H. Lloyd, *The Cutlery Trades*, London, 1913.

58   S. Pollard, *A History of Labor in Sheffield*, Sheffield, 1959, ch. 1; G. C. Holland, *The Vital Statistics of Sheffield*, Sheffield, 1843, pp. 107–17, 144–52.

59   *Sheffield Mercury*, 6 May 1837, 20 January 1839; *Figaro in Sheffield*, 16 January 1836; J. Baxter, 'Early Chartism and labour class struggle: South Yorkshire 1837–40', in S. Pollard and C. Holmes (eds.), *Essays in the Economic and Social History of South Yorkshire*, Sheffield, 1976, pp. 135–58.

60   A. D. Orange, 'The early history of the Yorkshire Philosophical Society: a chapter in the history of provincial science', unpublished Ph.D thesis, University of London, 1970.

61   R. K. Merton, *Sociological Ambivalence and Other Essays*, New York and London, 1976, p. 11.

62   See material in note 39; E. Robinson, 'The Derby Philosophical

Society', *Annals of Science*, 1953, **9**, 359–67; R. P. Sturges, 'Cultural life in Derby in the late eighteenth century', unpublished MA thesis, University of Loughborough, 1968; D. King-Hele, *Erasmus Darwin*, New York, 1963, pp. 13–46.

63 J. Alderson, *Address to Members of the Hull Subscription Library on Annual Lectures*, Hull, 1804, p. 15.

64 D. Fraser, *Urban Politics in Victorian England*, Leicester, 1976, pp. 11–14.

65 *Gore's Liverpool Advertiser*, 24 December 1835. Carson was made FRS in 1837.

66 Read, op. cit. (56), p. 131.

67 ibid., pp. 21–2.

68 Compare with Thackray, op. cit. (4), 672–82, 707–9.

69 Turner, op. cit. (41), pp. 2–6.

70 Turner, ibid., p. 2, and op. cit. (32), pp. 3, 5, 14; R. S. Watson, *The History of the Literary and Philosophical Society of Newcastle, 1793–1896*, London, 1897.

71 R. S. Porter, *The Making of Geology*, Cambridge, 1977, pp. 133, 147–8.

72 C. Harvie, *Scotland and Nationalism: Scottish Society and Politics, 1707–1977*, London, 1977, p. 129.

73 Porter, op. cit. (71), p. 155.

74 Smout, op. cit. (34), p. 475.

75 ibid., pp. 338–9, 350–2, 485.

76 Harvie, op. cit. (72), p. 135.

77 B. Robson, *Urban Growth, An Approach*, London, 1973, pp. 90–8.

78 Briggs, op. cit. (37), ch. 8; Read, op. cit. (56), pp. 50–2.

79 F. Sheppard, *London 1808–1870: The Infernal Wen*, London, 1971, pp. 318–19.

80 Thackray, op. cit. (25), 21, and op. cit. (4), 682; cf. A. Thackray and S. Shapin, 'Prosopography as a research tool in history of science: the British scientific community 1700–1900', *History of Science*, 1974, **12**, 1–28.

81 For varied approaches to the social dislocation of the period see R. Crompton and J. Gubbay, *Economy and Class Structure*, London, 1977, esp. ch. 4; G. Davidson, 'Explanations of urban radicalism: old themes and new histories', *Historical Studies*, 1978, **18**, 68–87; L. G. Johnson, *The Social Evolution of Industrial Britain*, Liverpool, 1963; Fraser, op. cit. (64); N. Smelser, *Social Change in the Industrial Revolution*, London, 1960; and, of course, Thompson, op. cit. (29).

82   As illustrated in works cited in notes 3, 4, 11, 12, 14, 15, 22, 29 and 36
      above.

83   For short accounts of which see E. J. Hobsbawm, *Industry and
      Empire*, London, 1968; H. Perkin, op. cit. (56).

84   Perkin, ibid., pp. 252–70; Neale, op. cit. (48), and his *Class and
      Ideology in the Nineteenth Century*, London, 1972; J. F. C. Harrison,
      The Early Victorians 1832–51, London, 1973, pp. 203–5.

85   R. E. Park, 'Human migration and the marginal man', *American
      Journal of Sociology*, 1927–8, **33**, 881–93; R. A. Schermerhorn,
      'Marginal man', in J. Gould and W. L. Klob, *A Dictionary of the
      Social Sciences*, Tavistock, 1964, pp. 406–7.

86   L. Wirth, 'Urbanism as a way of life', *American Journal of Sociology*,
      1938, **44**, 263.

87   Although at one point Smith comes very close to identifying the
      marginal man with the "outsider" (A. Smith, *Social Change, Social
      Theory and Historical Processes*, London, 1976, pp. 78–83), Merton's
      conception is of the individual who does perceive a social goal and
      whose social strategies are designed to achieve internalization. See
      his most recent statement in 'Insiders and outsiders: a chapter in the
      sociology of knowledge', *American Journal of Sociology*, 1972, **78**,
      9–47, and the very personal 'Structural analysis in sociology', in P. M.
      Blau (ed.), *Approaches to the Study of Social Structure*, London,
      1976, pp. 21–52.

88   This is, therefore, an analysis which has at its centre groups and
      individuals, and does not pertain to the fundamental social require-
      ments for the survival or destruction of the whole social system. See
      particularly D. Lockwood, 'Social integration and system integra-
      tion', in G. K. Zollsohan and W. Hirsh (eds.), *Explorations in Social
      Change*, London, 1964, pp. 244–56; N. Moizelis, 'Social and system
      integration: some reflections on a fundamental distinction', *British
      Journal of Sociology*, 1974, **25**, 395–409. For further discussion see
      Inkster, op. cit. (36), chs. 1, 4 and 6 and pp. 678–88; and I. Inkster,
      'Marginal men: aspects of the social role of the medical community in
      Sheffield 1790–1850', in J. Woodward and D. Richards (eds.), *Health
      Care and Popular Medicine in Nineteenth Century England*, London,
      1977, pp. 128–63.

89   For 'civic science' and later developments see R. H. Kargon, *Science
      in Victorian Manchester, Enterprise and Expertise*, Manchester, 1977;
      and for the pedagogic failure of popular science culture, I. Inkster,
      'The public lecture as an instrument of science education for adults –

the case of Great Britain 1750–1850', *Paedogogica Historica*, 1981, **20**, 85–112.

90 A. Briggs, *The Age of Improvement*, London, 1955; J. McLachlan, *Warrington Academy*, Manchester, 1943; U. Henriques, *Religious Toleration in England 1787–1833*, London, 1961; R. G. Cowherd, *The Politics of English Dissent*, London, 1959.

91 For a study of a range of social control mechanisms in one locality see C. Reid, 'Middle class values and working class culture in nineteenth century Sheffield – the pursuit of Respectability', in Pollard and Holmes (eds.), op. cit. (59); and of social control mechanisms generally, see A. P. Donajgrodzki (ed.), *Social Control in Nineteenth Century Britain*, London, 1977.

92 R. Williams, *The Long Revolution*, Harmondsworth, 1965, p. 164; M. D. Stephens and G. W. Roderick, 'Science, the working class and Mechanics' Institutes', *Annals of Science*, 1972, **29**, 353–9; J. Salt, 'The Sheffield hall of science', *The Vocational Aspect*, 1960, **12**, 37–43.

93 J. D. Bernal, *Science in History*, London, 1954, **i**, p. 665; Y. Ezrahi, 'The political resources of American science', *Science Studies*, 1971, **1**, 37.

94 M. Gowing, 'Science, technology and education: England in 1870', *Notes and Records of the Royal Society*, 1977, **33**, 71–90.

95 O. MacDonagh, 'Government industry and science in nineteenth century Britain: a particular study', *Historical Studies*, 1975, **16**, 503–17.

96 H. I. Sharlin, *The Convergent Century*, London, 1967. In addition such commentaries exaggerate the backwardness of British science and its divorce from industry. For an incisive and exhaustive counterbalance see M. Sanderson, *The Universities and British Industry 1850–1970*, London, 1972.

97 T. Kemp, *Historical Patterns of Industrialisation*, London, 1978, ch. 8; S. B. Saul, *The Myth of the Great Depression*, London, 1969; D. M. McCloskey, 'Did Victorian Britain fail?' *Economic History Review*, 1970, **23**, 446–59; R. C. Floud, *The British Machine Tool Industry 1850–1914*, Cambridge, 1976.

98 D. S. Landes, *The Unbound Prometheus*, Cambridge, 1969, pp. 235–6.

99 E. J. Hobsbawm, *Industry and Empire*, London, 1968, ch. 7.

100 A. Gerschenkron, *Economic Backwardness in Historical Perspective*, Cambridge, Mass., 1970.

101   I. Inkster, 'Meiji economic development in perspective: revisionist comments upon the industrial revolution in Japan', *The Developing Economies*, 1979, **17**, 45–68.
102   Kemp, op. cit. (97), p. 47.
103   E. J. Hobsbawm, 'From social history to the history of society', *Daedalus*, 1971, **100**, 20–45.
104   M. Crosland, 'History of science in a national context', *British Journal for the History of Science*, 1977, **10**, 95–113.

2 Whigs and savants: reflections on the reform movement in the Royal Society, 1830–48

Roy M. MacLeod

Introduction: the distorting mirror

Between 1846 and 1848, responding to overt pressure from within and the subtle threat of pressure from without, the Royal Society (henceforth RS) approved the most important alterations of its statutes ever considered in the 186 years of its history. These "reforms" were to have a dramatic effect on the self-image of the Society; they were also to have a profound effect on the public persona of the Society in its dealings with Parliament, with government and with men of science and other learned societies in London, the provinces and abroad.

For over thirty years, historians have seen this "age of reform" in the RS as an important part of the "professionalization" of science. Today, reading Lyons's semi-official account, the internal history of the RS since 1848 is presented as merely a chapter of consolidation, the natural consequence of enlightened policies set in motion by early Victorian "professionals".[1] This form of corporate self-description is not uncommon among institutions. Yet the historian's task is to expose the circumstances which such an account may unconsciously conceal.

In a rich comparative discussion of the "state of science" in nineteenth-century Europe, Merz drew attention long ago to allegations of "decline in science" in Britain between 1780 and 1815, and explored the prevalence of the amateur tradition in this country.[2] Historians have since continued to debate the merits of the declinist position, and its relevance to subsequent events in the organization of British science.[3] However, in 1951, Foote attempted to place these events in context, to understand the "great reforming energy" which accomplished this transformation, set against the background of reform provoked by the declinist attack, and the argument wielded by a small group of "scientific reformers".[4] He gave particular prominence to Charles Babbage,

and to his *Reflections* (1830), which criticized the RS, as John
Playfair and Charles Lyell had earlier criticized the universities for
failing to recognize and reward the cultivators of English science.[5]
Foote asserted that these events coincided with a time of 'reform in
practically every phase of English life'. He also argued that the
declinist attack had direct consequences for British science in two
ways: by challenging antiquated forms, it helped to strengthen the
Society; and by prompting the foundation of the British Association
for the Advancement of Science (henceforth BAAS), it opened the
way to improved relations between science and the state, and gave
new impulse to the efficient transformation of scientific ideas into
action.[6] In 1961 Williams located the climax of the declinist
campaign in the contested presidential election of 1830, when J. F.
W. Herschel was defeated by the Duke of Sussex.[7] Subsequently
Mendelsohn argued that these events were important principally
because in their wake emerged the BAAS, representing "professional" science.[8] None of these accounts considered the effect of
these events on the *subsequent* history of the RS. None was
concerned with the political context of the reform movement
outside the small band of "noisy ones" involved in the contest.
Nevertheless, the connotation of 1830, as a step towards professionalization has become received wisdom.[9]

Recent scholarship has shown the professionalization model to
be at best oversimple, and at worst untenable.[10] Indeed, the
professionalization model has impoverished our understanding of
the social order of science, ignoring actors' intentions, individual
interests, regional motives, and political expectations.[11] In fact the
motives, prejudices, and confusion surrounding the key events of
1830–48 convey far more than the emergence of a new professional
class dedicated to the "disinterested search for truth". Discovering
these factors has meant exploring what Cannon has called 'history
in depth'.[12] Rediscovering the RS, and its world of metropolitan
science, entails abandoning anachronistic distinctions between
traditional and modern, and between amateurs and professionals.[13]

The deepest difficulty in interpreting the history of institutions in
this period lies in untangling the complex relationship between
events in the RS and parallel events in British political life. During
these decades new mercantile, trading and professional interests
gained influence in the management of metropolitan science. Their
influence, giving way to pressure, eventually, for internal self-
management by these new interests, began in the 1820s and

culminated in 1848, the year which, for the RS, as for Europe, was the year of revolution. The events of 1830–48 also reveal the gradual and unexpected process by which traditional loyalties to Crown and Church were replaced by new contractual allegiances, or "hegemonies", based upon service to knowledge and utility to the State.[14] During this period the RS discovered, in Bagehot's phrase, the 'efficient secret' of the British constitution: a 'disguised republic', combining the dignity of monarchy and the 'representative power' of cabinet government.[15] Between 1830 and 1848, the RS experienced a version of the political and economic turbulence raging throughout Europe. Augustus de Morgan acknowledged that the 'Great epidemic which produced the French Revolution . . . showed its effect on the scientific world.'[16] The leadership of the RS, with strong ties to Crown and Parliament, was unable to ignore public charges of Toryism and favouritism, especially when such charges were supported by the interests of "reform" and "improvement" and wielded by a self-styled "philosophical party".[17]

The end of the Napoleonic Wars saw many traditional British institutions strengthened by the experience of those two decades. In common with restored Europe, British society in 1815 rested upon a pyramid of monarchical rule, assured of privileges ostensibly conferred by right and birth. In the institutions of English science, as in political institutions, the period between 1815–48 saw two great forces pitted against each other – men who owed allegiance to the traditional ruling classes, and men who looked to the new agencies of manufacturing and commercial power, and who made their bids for institutional power in the name of liberalism and specialized knowledge. However, these opposing interests were not always distinguishable. Few philosophers exemplified either extreme; most fell in between, and the RS was merely one alembic within which their interests were compounded.

Following the end of the war in 1815, Britain fell into severe recession. "Cheap government" became a watchword, drawing credit from the belief that the impoverishment of the country owed much to the incompetence of Tory governments, in power since 1800, and to the extravagance of the monarchy. Couched in class and party terms, this belief gained wide currency among middle-class liberals and working-class radicals. Economic and social distress contributed to political disquiet and social disorder. Civil peace could be had only at the price of electoral and administrative reform. Enfranchising the people would herald the end of sinecures and placemen, and the free air of representative government would cleanse the warrens of despotic institutions. The wars of 1793–1815, so the argument ran, had not been fought and won only to allow the British aristocracy to suppress liberalism and democracy at home.[18]

These circumstances affected British science beteen 1815 and 1850 in two significant ways. First, until Peel's government of 1841, there was little disposition to increase government spending (except through military or naval arrangements) on the pursuit of natural knowledge. Second, the institutions of science with aristocratic or royal connections were politically suspect; if they were staffed by court favourites, they were ripe for attack. Voluntary initiative and institutional self-help thus became not only corollaries of austerity; they were political common sense. Not until the spirit of reform reached the institutions of science would they be seen as fit recipients of public trust.

These economic and political factors, coupled with traditional intellectual rivalries with France, helped to shape the debate concerning the position of British science and the reform of the RS between 1830 and 1848. In these years, to borrow the felicitous phrase of Burn, Britain presents a 'distorting mirror', from which we receive conflicting images.[19] To resolve them, it is convenient to divide the reform period of the RS into three arbitrary phases: (1) from 1820 to about 1831, a period of apprehension; (2) from about 1830 to about 1836–8, a period of accommodation and reconciliation between the RS and different metropolitan and provincial interests; and (3) from 1836 to about 1849, a period of adjustment, in which the RS, succumbing to pressures for constitutional reform, acquired a new Benthamite image of philosophical integrity, public utility, open competition, and efficient administration. The first phase corresponds approximately with the discontented reign of George IV, from the riots surrounding the Six Acts to Catholic Emancipation, with the succession of Tory governments from Liverpool to Wellington; from the overtures of the declinists through Humphry Davy's tenure as PRS; from the death of Sir Joseph Banks to the election of the Duke of Sussex. The second phase coincides with the reign of William IV, the passing of the Reform Act, the short life of Grey's Whig Government, and subsequent party divisions between Peel and Melbourne; with the creation of the BAAS, the passing of reform legislation affecting both factories and poor relief, and the Duke of Sussex's tenure as PRS. The third phase begins with Peel's first administration, with early Victorian campaigns for administrative reform, and with constitutional reform in the RS. This phase coincides approximately with the Chartist agitation, with the elimination of protectionism, the repeal of the Corn Laws, and the succession of Melbourne's weak liberalism by the unsteady coalition of Lord John Russell. It ends with the arrival in Cambridge of Prince Albert as Chancellor, and the creation of the Natural Sciences Tripos, with electoral reform in the RS, the beginnings of the Great Exhibition, and the death of Peel. Throughout the period, the 'distorting mirror' reveals the RS following in both method and metaphor an uneven parallel in its transition from what Bagehot described as monarchical to parliamentary government, at the same time reflecting the new dispositions of political power in British society.[20]

I

Years of apprehension

In the history of the scientists' revolt, the turning point is by all accounts the contested presidential election of the Duke of Sussex in November 1830. That contest was the culmination of earlier attacks on the RS. During Sir Joseph Banks's courtly regime, the RS was assailed from several different sides.[21] The provincial towns, through their Literary and Philosophical Societies were creating important new traditions, effectively uniting mercantile, literary and philosophical interests.[22] Philosophical interests in London had advanced their own activities by creating new learned societies, which Banks opposed.[23] From 1820, the result was an undeclared "Thirty Years' War" in which the Astronomical and the Geological Societies, who shared the RS's premises in Somerset House, became its prominent rivals. Gradually, they helped to establish by 1850, in Gramsci's phrase, the counter-claims of the new 'organic' hegemony against the 'traditional' hegemony of Banks.

These claims were also being asserted by reforming interests within the Royal Institution (RI) and the medical profession. It would be surprising to find the RS unaffected by the RI's experience; both were caught simultaneously, as were the medical corporations, in cross-currents of public debate. Between 1800 and 1840, the RI played a central role in confirming a new social image of science and a model for its support.[24] The RI gave experience to men who later proved influential in the RS. Roget, an ambitious and versatile physician, and Brande, a favourite of Banks and a supreme civil administrator of science, were both recruited by Davy from the RI as Secretaries to the RS.

In preparing the ground for reform within the RS, the RI may also have provided a politically neutral ground for reformers and conservatives, a listening post for reverberations from public criticism, and an accessible and uncontentious practical body whose existence spared the RS from the necessity of giving popular practical interests special prominence.[25] Similar considerations arose in medicine. Medical men had always been prominent in the RS, but in the period 1800–48 their number (between 20 and 30 per cent of the Fellowship) and influence attracted particular notice.[26] By 1815 the Fellowship of the RS had become a mark of social status for medical men, particularly those of humble birth who aspired to the highest reaches of professional practice and social acceptance. Medical men on many sides of the reform movement, from Roget to Charles Bell, found the Fellowship an important social acquisition. To critics of the RS the persistence of "non-scientific" medical men was a perennial grievance. The three mystical initials FRS were ridiculed in translation as Fees Raised Since.[27] With the RI, the other learned societies and the medical corporations, the RS formed an interlocking directorate of metropolitan science. All reflected certain representative features of patronage and social place, and all represented a formidable interpenetration of scientific, cultural and careerist interests. In this sense, the reform movement in British science was embedded in the expectations of several different groups, all seeking to maintain status, influence and power.

In 1820 these expectations were still far from realization. The coincidence of the deaths of Banks and George III in 1820 released a fleet of political discontents upon an uncertain sea. Thus, Babbage wrote to Whewell:

... all sorts of plans, speculations and schemes are afloat, and all sorts of people, proper and improper are penetrated with the desire of wielding the sceptre of science. Whether this elective throne shall be filled by a philosopher or a peer, a priest or a prince, is a problem. . . .[28]

In 1820 the RS underwent its first presidential election for over forty years. Sir Humphry Davy was elected. He was acceptable both to philosophers and peers, and was devoted to preserving the delicate equilibrium of the society. Though Davy mended bridges with the sister societies previously ostracized by Banks, he altered little the role of President, preferring Banks's model of a benevolent monarchy of letters of which he was the titular head. His monarchy saw few sustained attempts at reform. Instead, the crown rested upon two "estates": rich men and men of rank who encouraged science by purse and patronage; and men of science who were thus patronized. Successful government depended upon the harmonious co-operation of both, secured by good administration. Davy's administration was conducted largely through the efforts of Brande, as Senior Secretary, and Davies Gilbert, as Treasurer. In 1824 Brande was joined by Herschel, who served until 1827. That year, however, Herschel and Gilbert collided, and their collision affected the RS's future for the next two decades.

In 1827, aged 35, Herschel was one of the most celebrated philosophers of his day. Elected to the RS in 1813, he became a founder member and officer of the Royal Astronomical Society in 1820. Since Cambridge, with Babbage, Peacock and Whewell, he remained a central figure in a cultural network whose influence pervaded English science and letters.[29] Politically and philosophically, his sentiments were liberal, though not Whiggish, and certainly neither radical nor utilitarian. His life had been unmarked by religious or civil turbulence; he was too young to remember Robespierre. Gilbert, aged 60, in 1827, had little in common with his younger colleague.[30] He bore memories of the anti-Jacobin mob burning Priestley's house in Birmingham. From 1806 to 1832, he was Tory MP for Bodmin, and an admirer of the conservatism of Fox and Burke. In 1810 he had warned in Parliament that the tumultuous storm of democracy would lead to a gulf of despotism. In the 1820s he spoke for public order in the face of Peterloo, and against Catholic emancipation in the face of his party.

In 1827 events at Westminster and in the RS followed a parallel course. Electoral and administrative reform were the dispositive issues. In February 1827, Liverpool's health destroyed, Canning formed a government held together only on mild reforming lines. Gilbert saw himself caught between the horns of corruption and revolution. The same dilemma confronted the RS. Gilbert's fears of political reform outside the RS were fanned by fears of 'mob rule' within, of 'disorder leading ultimately to dictatorship'.[31] It followed that, like Parliament, the institutions of science wanted only an alteration or two. Like a monarch, the President could, by tradition, hold the chair until death or abdication. Councils were coteries of friends, selected by the President; financial accounts were closed except to him and the Treasurer. Papers were read infrequently, and only by the President's permission, and publication was slow. Fellowship elections were casual affairs, conducted at ordinary meetings and often on the nomination of the President. The new Royal Medals, awarded from 1826, were virtually in the gift of the officers.[32]

In March 1827, opposition to the RS's courtliness came to a head when Gilbert received a demand from a small group of Fellows to publish a statement of property belonging to the Society. Subsequently, a committee of Fellows, including Herschel, Babbage and Sir James South, was asked to report on Fellowship elections and the appointment of officers. Its report recommended a ceiling of 400 Fellows; elections restricted to four new Fellows each year, to reduce purely social interests; and a more powerful Council, empowered to select its successors and to oversee finance.

On Davy's retirement in the summer of 1827, Gilbert was invited to be President. Neither he nor Davy saw the selection of a President as different from the nomination of a candidate to a pocket borough. But Gilbert's response was tempered by recent events. With Davy's support, Gilbert called upon Sir Robert Peel to be President. The danger Gilbert saw was of outrageous republicanism. 'The Great Contest that I allude to,' he told Peel, 'is the Conflict of Aristocratic and Democratic Power. I wish the Royal Society rescued from the latter.'[33] Peel declined the invitation. In his stead, Gilbert was elected President in November 1827. In December 1827 he squashed the Fellowship Committee's report without further discussion. The result was only to drive discontent underground. In 1827 Herschel resigned as Secretary, resolved to fight against the injustices he had seen.[34]

In the presidential election of Sussex in 1830, three issues linked philosophy and politics. First, in 1829 three leading statesmen of the *ancien régime* of English science, Wollaston, Young and Davy, had just died. Into this vacuum rushed speculation about the "new men" who might succeed them. Second, simmering discontent with the financial and electoral administration of the Society came to the boil. Third, the Society's proximity to medical politics threatened to convert it 'into an engine for party purpose and self-interest'.[35] It was even rumoured that Henry Warburton had been offered the Presidency in return for his support of a successful Anatomy Bill.

Gilbert feared that the power of the old order in science was passing, as part of the process which was transferring political power in the State from the aristocracy to the middle classes. Both as an MP and PRS, he tried to soften the impact of this process. As his Secretary, he chose Roget, a close associate of leading utilitarians. But his efforts at accommodation were to no avail. Gilbert and Roget suffered steady rebuke from Herschel, Babbage, South, and correspondents in *The Times*.[36] The public was treated to the spectacle of South, President of the Astronomical Society, accusing Gilbert, PRS, of perfidy and incompetence. With the publication of Babbage's *Reflections* in May 1830, all hopes of peace were lost. Moreover, the RS could not free itself from the web of intrigue among the medical corporations. In April one FRS used *The Lancet* to describe the abuse of the Society by those whose fashionable practices rendered it 'a medical advertising office, a very puff shop for the chaff of medical scribblers'.[37] To such medical men, the absence of reform in the RS was doing irreparable harm to the reform of English medicine.

Faced by a constitutional crisis, Gilbert looked for a safe successor. His purpose was to secure for the Society the continuance of aristocratic interest, liberalized in practice, serenely removed from the disquiets of civil war. By August 1830 he had approached the Duke of Sussex, whose well-known hospitality at Kensington Palace assured continued yet liberal royal patronage and a plausible safeguard against radicalism in medical and scientific affairs.[38] In the event, his strategy sadly underestimated the participatory sentiments of a substantial number of London Fellows.

In early November 1830, a Requisition signed by thirty-two Fellows was sent to Gilbert demanding the publication of all correspondence concerning his own plans and the negotiations with Sussex.[39] This procedure was unprecedented; all Gilbert's fears of the *vox populi* were apparently being realized. Alarmed by this combination, Gilbert explained to a meeting the circumstances of his invitation to Sussex. When he had finished, the meeting passed two resolutions. The first asked that in future officers should be selected for 'acquaintance with the conditions and interests of Science'. The second, moved by Herschel and seconded by Faraday, asked that the Fellowship should choose its future officers, albeit from a list drawn up by the President and Council. The meeting ended with a vote asking Sussex to retire.

The resolutions of this unprecedented meeting were by nature purificatory; neither implied a revolutionary break with the statutes, but both broke sharply with custom, at a time when customary forms were under great pressure. The country itself had just emerged from a bitter general election in July and August. Throughout the country, middle-class radicals were pressing for reform, uniting with such working-class interests as the Radical Reform Association and the British Association for the Promotion of Cooperative Knowledge. The autumn saw reports of rick-burning and Swing's signature in the South, and a flurry of unstamped papers, demanding concessions to working-class interests.[40] Gilbert personally associated the Fellows' request with the firing of hayricks on his Sussex estate that autumn. Demands for reform did not require the July Revolution to spark off calls for a new Government, but November saw the arrival of Grey's administration – the first Whig Government in decades, with leaders in political, medical and educational reform (including Brougham, the new Lord Chancellor) holding high office.

The same month, Wellington issued his famous refusal to countenance any measure of reform. Party lines at Westminster seemed mirrored in the battle lines at Somerset House. But the Requisitionists signalled neither revolution nor a victory for reform. The picture is more complicated, reflecting only the complex nature of political alignments. The Council (including many of the rebels) was itself divided on the subject of patronage. Faraday, no radical in politics, disappeared from the discussion. Most of the Fellowship remained aloof. By whom should Gilbert have been bound? Under these circumstances, the announcement of Herschel's candidature against Sussex forced a choice. Gilbert was prepared to accept constitutional alteration, while at the same time he rejected the concept of an exclusively "philosopher's Parliament". Such a change would destroy the social basis upon which the Society depended.

At the next meeting of the Council on 18 November, Gilbert procedurally disallowed the resolutions passed at the unconstitutional meeting of 11 November, and announced his intention of resigning. This effectively mobilized opposition. Spurred by Babbage, a declaration of support for Herschel was circulated. This important document reveals that of the eighty signatories at least thirty-three were not "scientists". Forty-five were not authors in *Philosophical Transactions*, although twelve of these – Lyell for one – were certainly considered "scientific". Some of the eighty represented reforming parliamentary interests (both among Tories and Whigs)[41] and several represented legal reform or utilitarian sympathies.[42] Three (Brodie, Green and Warburton) had been active in reform in the Royal College of Surgeons since the 1820s, while, of the physicians, four (Chambers, Holland, Prout and Elliotson) were prominent members of the medical establishment. Significantly, the Declarationists included experienced Fellows, who saw the constitutional integrity of the Society as the central issue. Of the sixteen past or present Councillors who signed, seven were actually on Gilbert's Council at this time. There was no distinct separation of opposing party interests. Barlow, Ellis and Sedgwick, for example, supported Herschel, but were not in Herschel's prospective cabinet.[43] In fact, Barlow and Ellis, although they ostensibly supported Herschel, actually remained on Sussex's Council after 1830! Their concern was that shared by many metropolitan improving interests. Under the circumstances, it is not surprising that the list contained few titled gentlemen, or clergy-men, or for that matter, Fellows resident outside London.

The Herschel Declaration demonstrated the diversity of opinion within the Society; compromise was required, but capitulation would have been disastrous. In any case, Gilbert (and Sussex) had the support of over 400 Fellows, including substantial representatives of the Church, the Army, the Navy, and much of the legal and medical establishment, who did not sign. To alienate these elements would have, in Gilbert's eyes, reduced the influence of the Society in the affairs of the nation.

Meanwhile, in a small book entitled *Science without a head*, Augustus Granville imparted a cool perspective to the contest.[44] His survey of the 651 current Fellows argued that only one in five could be regarded as an active philosopher: most were merely *savants en credit*. His sharpest arrows were reserved for his medical confreres (fifty-five physicians and eleven surgeons) whose motives for earning the FRS had little to do with useful inquiry. Superficially, Granville confirmed the justice of Babbage's attack. But there were two important differences. Granville's book was also a treatise on political sovereignty. He rejected the suggestion that science was in decline and made no claim for science as a profession. Instead he argued for the replacement of inefficient privilege by efficient patronage, regulated by a new constitution. Second, Granville believed the reform of leadership lay not in the choice of an enlightened politician or even a good astronomer: the first committed the Society to definite political alignments; the second had merit, but other qualifications were essential in a President. In effect, Granville's book parried the "noisy men" with a spirited defence of the constitution (and *ipso facto*, Gilbert) and concluded by supporting the election of Sussex. It is difficult to gauge Granville's influence on the majority of Fellows. Some in the liberal camp attempted to dissuade Sussex.[45] For reasons which remain obscure, Sussex declined to withdraw. Perhaps like Gilbert, he feared the prospect of victory by a reforming movement, led by petty sectional interests, to which Herschel had been drawn.

The famous poll on St Andrew's Day, 1830, was 119 to 111, narrowly in favour of Sussex. His narrow victory would have been wider had more of the 600 Fellows voted. The outcome was to Gilbert's manifest relief. At the same time, historians have interpreted it as a loss for the reforming party. As Williams put it, 'the cause of reform had been soundly beaten, and the reformers retired in disgust'.[46] Even *The Times* lamented that 'the first scientific establishment in the Empire has obtained a Prince and missed a Philosopher for its President'. But the question was not a simple gloss on the principle of peers versus the people. The outcome challenged the Society to find a political model which would successfully govern English science in a period of rapidly shifting allegiances. Gilbert rejoiced: 'I have retreated over a Gold Bridge with Purple Banners.'[47] But he knew that the work of adjustment had just begun.

II

Years of adjustment, 1830–8

For the next eight years, Sussex, aided by Gilbert, tried to create a *rapprochement* between the RS and the political life of the country. This was made easier by the fact that, like the institutions of State, the Society continued largely unaltered in its ways. Fellowship elections continued in what Granville called 'a desultory fashion' at every ordinary weekly meeting where a quorum of twenty-one was present.[48] Herschel left the country to begin his observatory at the Cape of Good Hope; Babbage, Murchison and Sedgwick turned to the new BAAS. In 1832 Babbage took an unsuccessful excursion into national politics. His electioneering play, 'Politics and poetry, or the decline of science', eloquently conveyed his belief, with Gilbert, that the two were really one.[49]

The 1830s witnessed an expansion in local scientific activity, led by the BAAS. As in the country's political life of the last decade, the place of London in the "strategy of agitation" had changed; the provinces, and provincial scientific societies supplied much of the energy behind this new activity.[50] Under Sussex, the RS quickly accommodated itself to the BAAS's existence, while in fact accepting little internal change. Moreover, Sussex began to cultivate the Fellowship. His combination of political wisdom and intellectual taste was vital; for the years 1830–8 were a period of great contention within the Society.

It was no less a period of contention in the country, with four general elections occurring in the space of ten years. Both in Parliament and the RS, Sussex supported the 'most that could be pushed, and the least that would satisfy the country . . . '.[51] At his first meeting in December 1830, he announced his intention to hold soirées on Monday in the season, to encourage 'a familiar and useful intercourse of wealth and talent of men of rank and men of genius and other grandees of society'. In the uneasy calm following the Reform Bill, nothing could be 'better calculated', said the *Gentleman's Magazine*, to promote harmony and good feeling between the 'three Estates of the Kingdom'.[52] To share power, more Vice-Presidents (an average of six per year instead of two) were created, and were selected from old friends and adversaries alike. Opposition was accommodated in the Council. Of the eighty declared supporters of Herschel, eleven became Council members within two years, and a further eight, within the next decade.

In the meantime, the major work of administrative reform was shelved. This was, ironically, made easier by the fact that during the years 1832–5, the country, and much of the Fellowship, were preoccupied with parliamentary reform (1832), factory and poor law legislation (1833–4) and with ecclesiastical, educational and medical issues.[53] Within the medical world, the cholera outbreak of 1831 forced attention to the inadequacy of official medical arrangements.[54] In their protest against the Royal Colleges, Thomas Wakley and Charles Hastings pursued a model similar to that of the British Association, in forming the Provincial and Medical Surgical Association at Worcester in July 1832.[55]

Between 1834 and 1835, following Gilbert's retirement, Sussex's Council contained several of these reformist surgeons, while other figures, including Peel, improved the public image of the Society and of English science.[56] Peel's brief term as Prime Minister between 1834 and 1835 saw a baronetcy for John Barrow in 1835, and civil list pensions for Georgy Airy (who became Astronomer Royal), Mrs Somerville and Faraday. In 1836 the pensions earlier awarded to Brewster, Dalton and Ivory were increased.[57] However, Peel's labours were interrupted by Melbourne's Whig Government, which, for six years between 1835 and 1841, generally did little to encourage science.[58] But Sussex, working through the Reform Club and the Lords, mediated between Melbourne and Warburton's parliamentary radicals, and managed to keep the Government's sympathy. It was ostensibly at his request that Herschel, his former opponent, was given a baronetcy in the Queen's Coronation Honours in 1838.[59]

The years between 1830 and 1837 also saw a new alliance between the cultivators of science and wider commercial, economic and political interests. In 1831 James Douglas, the political economist, claimed that philosophers required

the union of the patronage of the government with the interest taken in its prosperity by the nation at large. This interest could only be universally diffused by a general and voluntary society.[60]

What had begun as a protest against the inadequacy of royal patronage had grown into a debate about the sources of patronage generally. The remedy, Burke once said, lay in association. In fact, this role soon became the special task of the BAAS.[61] The radical image of the BAAS at first had proved startling, especially when associated with the provinces.[62] Recalling the decade 1830–40, Babbage saw that science 'like other kinds of enterprise...has resorted to association for the furtherance of its object and for the support of its expenses to the contributions of its associates'.[63] The BAAS served the RS's interests in three ways. First, it offered a convenient public forum for the promotion of science without at the same time compromising the RS. This took two forms. Some public concession was desirable politically. As Lyell wrote to Darwin in 1838:

... in this country, no importance is attached to any body of men who do not make occasional demonstrations of their strength in public meetings ... nothing is to be got in the way of homage or influence, or even a fair share of power, without agitation.[64]

The BAAS gave philosophers (including FRSs) a chance to turn public orators at no direct cost and at considerable indirect gain to the Society. Moreover, the Fellows of the RS could work, if necessary, through the BAAS in seeking government support for tidal observations (1834), an Antarctic expedition and magnetic observatories (1839). The presence of claims launched by *two* organizations, one avowedly committed to self-help, ensured a better hearing.

Second, the BAAS offered a timely concession to men who desired the status of philosophers without implying the social sanctions that the letters FRS had hitherto bestowed. As Babbage wrote in 1832,

It is highly probable that in the next generation the race of scientific men in England will spring from a class of persons altogether different from that which has hitherto supplied them.[65]

The BAAS's existence relieved the RS from direct responsibility for the democratic consequences such a development might imply.

Finally, the BAAS, widely reviewed in the major weeklies of London, deflected gossip away from the metropolis itself, while providing the RS with a social laboratory in which to test national feeling. Borrowing parliamentary language, the BAAS helped the RS restore itself as an upper house of science, aided now by a substantial Commons. What Tyndall called 'the great travelling congress' could work together with the 'old stationary palladium of British science'.[66]

In the late 1830s, it was perfectly plausible for a visiting savant to write that 'England is not the land of science – es existirt dorten nur ein Weitgetriebener Dilettantismus'.[67] It was plausible, but not quite accurate: many undercurrents were disturbing the superficial calm of the English scientific establishments. Far from disappearing, complaints about the "state of science" revived. Bulwer Lytton's influential *England and the English* (1833), which contained a chapter strongly suggestive of Babbage's influence, revived the twin questions of patronage and recognition. 'Science is not higher on the continent than with us,' the argument went, 'but being more honoured, it is generally more cultivated.'[68] While Britain led in manufacturing and practical science, the 'circle of speculative science' was narrow and confined. According to Lytton, this deficiency could only be made good by endowing science at the ancient English universities, and by bestowing public honours on its devotees. As Granville argued in his second book, *The Royal*

Society in the Nineteenth Century (1836), the required remedy was fundamentally political. The existing system of patronage was inimical to the philosophical interests of a wider franchise. Remedy required that patronage by rank be displaced by the patronage of fellow philosophers.

In 1838 Sussex resigned. In his last presidential address, he had welcomed Herschel, whose return could well have opened an old wound. Peel, still out of office, was canvassed, but again declined, asking that 'the chair should be filled by some distinguished man who had devoted his time and faculties to some branch or other of science'.[69] Eight years before, the election for the Presidency was hotly contested. Now, tactfully avoiding interference with 'the free use of the franchise which every Fellow possesses and is expected to ... exercise',[70] Sussex welcomed the Council's nomination of the Marquess of Northampton, a man 'without more pretensions to science or philosophy than a good education and a taste for such studies would naturally justify'.[71] Northampton, aged 48, a graduate of Trinity College, Cambridge, a former Whig MP (1812–20), an FRS since 1830, and a mineralogist of note, was easily underrated. For the previous two years, he had been among the reform party in the Council, and was a favourite compromise candidate. Through him, achieving recognition for "philosophical patronage" became the dominant theme in the next phase of reform.

III

Years of accommodation, 1838–49

By 1849 the RS had completed the third and final instalment of its internal reform. A decade which culminated in revolution in Europe, witnessed in Britain the collapse of Chartism, the beginnings of reform in the civil service, in public health and the universities, and recognition of the RS as a responsible public body.[72] At Westminster, Peel's administration fought through a programme based on sound finance, efficient administration and social legislation. In analogous fashion, Lord Northampton, similarly encumbered by an unreformed civil service managed by his Senior Secretary, Roget, guided the RS towards electoral reform, and to the recognition of new "philosophical" constituencies within the Fellowship and the Council.

Returning to government in September 1841, Peel watched the

"philosophical" movement grow in influence. Though aware of the odious power which patronage confers, Peel nevertheless used his patronage to encourage the process. Peel's efforts reached beyond pensions to grants for scientific purposes, and knighthoods for scientific reformers.[73] Yet Peel's support of science and the Society could not alone restore the Society's image. In 1842, Augustus de Morgan, never an FRS, claimed that

The RS is the focus of aristocratic science and scientific aristocracy.... It has justly survived not merely by the labours of those who cultivated science, but also by the financial support and promotion of those who did not. If it were suggested, however, that the RS had a public responsibility, as well as a right to service, it was not clear that it had done its duty.[74]

By the mid 1840s, despite Sussex's efforts, the RS was neither completely 'dignified' nor 'efficient'.

When Victoria ascended the throne in 1837, the monarchy was at a low ebb. Like Victoria, Northampton had the task of restoring lustre to the sceptre of English science. Steps were taken to sustain the *entente* with the BAAS and to improve diplomatic relations with those learned societies which had become what one observer called 'antechambers to the Royal'.[75] In his first address, Northampton announced the launching of Ross's expedition to the Antarctic, pressed on the Government by the BAAS, and the programme of colonial magnetical observations organized by Sabine. Continuing Sussex's tradition of soirées, he similarly pursued Sussex's Whiggish interests in encouraging "scientific utility" and in improving the *Philosophical Transactions*.

For fifteen years, thanks to the conciliatory posture of Sussex and Northampton, the Society's Council Minutes were silent on the subject of constitutional reform. Like Sussex, Northampton avoided being carried away by extremists of either group.[76] As he wrote to Granville 'You see, my dear doctor, that we are accepting little by little your suggestions; all the rest will come by and by.'[77] In fact, by 1847, Northampton realized that he was virtually controlled 'by what is called *his* Council'. Increasingly, the Council, and not Northampton, was held responsible for the Society's actions – to contain radical sentiment, to curtail professional self-advertisement, and to eliminate favouritism. As the reformers of 1830 had shown, these reforms depended upon modifications in the practice of Fellowship elections, in the selection of the Council, and in the tenure of the President. To check the indiscriminate admission of new Fellows, Northampton re-established proposal forms on which grounds for nomination were required.[78] The issue of presidential tenure came next. In 1848, after a decade in which, by general consent, he gave more general satisfaction to the Society than any President since Banks, Northampton recognized that no future President could ever again reign so long, lest he become 'too absolute – nay jobbing – or at all events, if he lives to be old, he is likely to become superannuated, and not to know it'.[79] By 1848, however, these sentiments were overtaken by more sweeping proposals which had slowly gathered impetus in the preceding three years.

In May 1846 a debate arose among some of the Fellows over the award of one of the previous year's Royal Medals, to T. S. Beck, the physiologist, for a paper on the nerves of the uterus. Contrary to custom, Beck's paper had not been previously presented to the Society or published in the *Philosophical Transactions*. As usual, the President, advised by the Council, had chosen the winners. When Beck's award was announced, Dr Robert Lee, FRCP, an obstetric physician and former Council member, claimed that his own work on the same subject had been overlooked and then, after a cursory reading, rejected by the Council. Lee was ultimately overruled; but the radical medical press, and the medical community generally, were enraged.[80] Immediately, the charges precipitated the dissolution of the Committee of Physiology; eventually, they were to prompt Northampton's retirement.[81] Coupled with a threat of legal proceedings they ultimately forced Roget's resignation. The issues of principle involved were seized upon by William (later Sir William) Grove, who was to play a central part in the main events of the next three years. Grove, then aged 34, an Oxford graduate and patent lawyer, was elected FRS in 1840, and was on the Council from 1845. In 1846 he published his best known work on the *Correlation of Physical Forces*. To the reformers, his combination of legal and philosophical talents proved decisive. Shortly after the Beck affair, Grove took steps to mobilize a charter committee of the Council to consider alterations in the Society's constitution. The political symbolism of the Chartists was not lost on him or his contemporaries.[82] Before the summer recess of 1846, with little public display but with great determination, the charter committee recommended 'organic changes of the most comprehensive kind'.[83] The committee recommended a triennial term for the President, more efficient organization of the subject committees, and most important, the restriction of new Fellows to fifteen each year, chosen by ballot. Irritant sand had produced a reformist's gem.

Since 1830, the Fellowship had broadly accepted the argument that its President should be a man of position, preferably with access to the Crown, Parliament, and departments of State. But the internal government of the Society was a different matter, requiring a new form of ministerial responsibility. Great significance, therefore, attached to the committee's recommendations, which implicitly redefined the Council, in parliamentary language, as a cabinet, and the President, by implication, as no more than its prime minister, and to that policy which would open the gate of Fellowship to men of lesser means but greater philosophical ability.

Throughout the autumn and winter of 1846–7, these issues became prominent in those sections of the medical press which spoke for the corporate conscience of medical reform. *The Lancet* aimed an attack upon Roget, whose twenty years in office had, in its view, blocked the representation of new interests. These charges, of great importance to medical FRSs, came to a head at a Special General Meeting of the Society, held on 11 February 1847, specifically to consider the proprieties of the Beck affair. At a time when the Royal College of Surgeons was contemplating the election of Fellows by examination, the recognition afforded by Royal Medals was vital to medical standing.[84]

Meanwhile, in early 1847, appeared an anonymous publication, signed 'FRS', entitled *Thoughts on the Degradation of Science in England*. The author, later alleged to be Charles Babbage, revived the rumours of decline in British science, and attacked the Society's administration, which, in his view, had since turned this decline into calculated degradation. The tract was far more political in its overtones than Babbage's book of 1830, but despite its rebarbative prose, it aroused far less debate. In fact, the Council had already begun to respond, although its response was muted: nothing was done to remedy the failure of the sectional committees, whose comparison with those of the BAAS invited scorn; and for legal reasons which are not fully revealed in the Council's Minutes, the recommended changes in the charter and statutes were dropped, save for those concerning Fellowship elections. In the event, following legal advice from the Law Officers of the crown, this single proposal was approved by the Council on the Society's behalf, on 10 February 1847. The equivalent of electoral reform had been achieved, at last, and apparently without revolution. Were it not for the medical presence, it might have been silent as well.

None the less, the result of this simple change was far-reaching. Given the age structure of the Society, the elective limitation to fifteen Fellows immediately reduced the size of the Fellowship. Where there were 764 Ordinary Fellows in 1847, there were only 630 in 1860, of whom 330 were said to have 'higher scientific qualifications'.[85] The RS was the only major learned society in England to shrink in size in this period; and within two decades, there was a formidable queue of eligible men, including a large number of disappointed physicians.[86] To satisfy sceptics, important features remained unaltered. The statutes permitted, for example, any number of the Fellowship to be elected in certain privileged classes based on rank; and the Society remained highly medical, about 20–25 per cent of all Fellows elected between 1848–1900. The Society's finances suffered, but not catastrophically. Moreover, the reputation of the Society vaulted. The list of fifteen names each year, as Granville said, enabled anyone to 'judge of the real worth of our present standing in the opinion of the scientific world'.[87]

Reflecting the Liberal Government of Lord Russell which had replaced Peel in 1846, the Council which had introduced these changes bore the timely complexion of a coalition cabinet.[88] Its political sympathies were conciliatory, if not quite reformist; and favourable to the mixture of political sympathies which combined in what might be called the "philosophical" party. Thus the events of the following few months must be understood not as an assault against the Society but as an attempt by a coalition cabinet to consolidate its policies.

With the election procedures altered, the philosophical integrity of the Society could be strengthened. To air demands for these further reforms, a ginger group, representing a broad base of philosophical interests, was well advanced by February 1847, when J. P. Gassiot and Grove proposed the creation of a new society within the Society. In April 1847 twenty-seven Fellows met at a London hotel to form the Philosophical Club. It was to meet monthly on Thursdays at 5.30 p.m., just before the meeting of the RS, its purpose being to 'promote the scientific interests of the RS [*sic*] and to facilitate intercourse between fellows cultivating different branches of natural science'. To honour the year of its foundation, membership was limited to forty-seven Fellows, who were required to have contributed a scientific paper to one of the metropolitan learned societies (a rule inspired by the BAAS).[89]

The Club bore a close relation to the coalition cabinet of the RS. Superficially it appeared a creature of that cabinet, representing key metropolitan and professional interests.[90] Grove was appointed Treasurer, and Edward Forbes, Thomas Graham, Leonard Horner and John Royle formed a Committee of Management. There was to be no President, merely a chairman to guide discussion. But a substantial number had experience of government.[91] Despite the presence of a strong medical interest, the Club's direction was deliberately philosophical rather than professional: those elected were mainly men who already earned their living by means other than the cultivation of science. Almost immediately they began to influence public opinion, not by fostering political cabals, but by strengthening the position of the RS among the learned societies of London.

With the creation of this philosophical vanguard, the RS had the 'elements of a great organisation . . .'. 'Let [its Fellows]', urged an FRS in *The Athenaeum*, 'by some means seek a reunion with the cultivators of the natural sciences which they have lost.'[92] The two issues to which the Club gave immediate prominence were the unification of the different scientific societies in one locality, preferably in one building; and the normalization of relations with the other societies. The first object it pursued for over two years, leading eventually to the acquisition of Burlington House in 1852; the second, it proposed for a time to achieve by inviting the Linnean, the Geological, the Astronomical and the Chemical Societies to send their best papers to the *Philosophical Transactions*. Both aims indicated the Society's new will to leadership.

Within the Society, the fuse lit by the Beck affair was kept alight. Gradually, the RS had moved towards open government through the publication of accounts, rotation of committees, restriction of members, election by merit, and had nearly limited the tenure of its officers. But these policies were informal, and could still be overturned by an unsympathetic President or Secretary. At a heated Council meeting on 9 February 1848, Sir Henry De la Beche advanced, and then, on reflection, withdrew a proposal to restrict the office of President to two years, which would have reduced the President to the status of merely a superior member of Council.[93] But what would be the future principle of government in the Society – monarchical or parliamentary, aristocratic or philosophical; in Bagehot's phrase, 'dignified' or 'efficient'?

Early in the revolutionary year of 1848, Northampton indicated his wish to resign. His succession unavoidably again brought into focus the constitutional role of the officers. In March 1848 *Punch* supplied an advertisement for a new President:

Wanted, a Nobleman who will undertake to dispense once a month, upon rather a liberal scale, tea, lemonade, and biscuits, for a large assembly. The company is select, and he will be allowed to mix with some of the greatest men in England. Sealed tenders, stating most liberal terms, to be sent in to the Royal Society, marked 'President'. No scientific or literary men need apply.[94]

As if to anticipate criticism, Northampton proposed the Earl of Rosse, the astronomer. Herschel, again proposed, was weary of the fray, and willing to defer to Rosse. There were few other possibilities. Edward Sabine, an "artful dodger" perhaps, but certainly an astute politician, again (unsuccessfully) suggested Peel. Babbage had protested against an aristocrat, but the election went the President's way. Rosse, who believed in continuing an element of Tory paternalism in the Society's government, offered the best compromise, though his residence in Ireland for most of the year hardly promised a major social force in London.[95]

As the election drew near, others urged the replacement of Roget to remedy the 'negligent manner in which the business of the Society has been conducted for years past'.[96] In May, *The Athenaeum* speculated that both Roget and Northampton might soon retire, providing an opportunity to complete the Society's reform.[97] In fact, in what amounted to a general election, the Council was captured by the progressive movement in the Society, including Airy, Horner, Gassiot, Lyell, John Phillips, Grove and Wheatstone. The Philosophical Club had revealed its new-found strength. Altogether, thirteen members of the Council belonged to the Club. But the Secretaryship was not to be decided by reforming interests alone. The time had come when, following the electoral reform of 1847, new constituency boundaries had to be defined. Increasingly, they would be defined in favour of the philosophical specialities. The medical fraternity, for example, claimed Roget's post as their own by right. As W. B. Carpenter wrote to Grove: 'If battle is not between the Reforming and Conservative parties, but between the physicists and the naturalists, we must support our own sciences.'[98] And Walter White, the RS's Assistant Secretary, asked perceptively, would 'science or faction conquer'?[99]

Following a heavily contested election, Thomas Bell (dental surgeon and Professor of zoology, King's College, London) became the new Junior Secretary; and in January 1849, the Council, under Bell, began the work of detailed administrative reforms long deferred under Roget. Seven new sectional committees were created, each consisting of between twelve and twenty specialists to consider papers, and to make recommendations for the Copley and Rumford Medals, and for the use of the Donation Fund. Hereafter, administrative decisions would never be far from philosophical interests. Would science or faction conquer? Ironically, in the future they were indistinguishable.

The election on St Andrew's Day, 1848, was a significant turning point in the Society's history – not because it heralded the advent of a scientific liberalism based on reform, but because it replaced conservative principles based on patronage with principles based on disciplinary prerogative. Hereafter, the Society would be both 'dignified' and 'efficient'. There exists no better index of its improved public reputation than the good opinion of men of all parties, and particularly, of Russell's Liberal Government, which had watched closely the events of 1848. In October 1849, Russell offered the Society an unprecedented parliamentary grant 'for the promotion of science and the encouragement of scientific men'.[100] This bespoke not only the confidence of a Liberal Government; it was tacit recognition that the Society had set itself in order.

Conclusion

By 1850 the Society had repaired its public and philosophical image.[101] After twenty years, science was no longer without a head. The Reverend Abraham Hume remarked that this was a 'pleasing set off to the well known observations, both correct and important' respecting the decline of science in England.[102]

From the reform decades of 1830–48, several considerations arise. This essay has suggested first, that the events of 1830 were not, as Williams has written, 'a victory of the amateurs'; it was, instead, like the Reform Bill of 1832, a victory for moderation.[103] Equally, the "triumph" of 1848 was not so much a victory for Whiggism, as a victory for coalition, conservatism and compromise – ultimately for a Peelite strategy of assimilation and accommodation. In this context, the language of politics becomes more than a matter of metaphor. Parliamentary representation, electoral reform, a self-determining extended franchise – these were the issues which dominated England and which were reflected in all her major institutions. The institutions of science were not excepted. As *The Times* perceptively observed:

The Royal Society, like all other institutions, must take account of the fact that democratic principles now govern the world in all fully civilised communities – the chief scientific society must in all matters itself be scientific.[104]

In Bagehot's terms, the government of the Royal Society moved from absolute to constitutional monarchy. Within another generation, the transformation was complete; the President would become, not a peer, but a prime minister, *primus inter pares*, presiding over a cabinet rather than a *curia regia* of Council members. Within two decades it was agreed to keep presidential tenure to the same limits which, by custom, each British government observed. And no longer could cabinet officers be appointed without regard to the constituency interests representative of specialized disciplines.

It should now be clear that the application of a simplified model of professionalization has honoured the events of 1830–47 more in the breach than in the observance. It is far from clear that the 'uprising of scientists at the RS in 1830 was in part a proclamation that science could now stand on its own, that it had the prestige of a profession'.[105] Such an assessment depends too much upon the accounts of Babbage and Brewster; it also distracts the reader from the historical context within which the uprising occurred. That context was as much political as professional, as much concerned with the parochial concerns of a traditional elite, fearful of abrupt change, as with the intellectual interests of an organic elite seeking recognition.[106] This becomes all the more clear when the fearful events of 1830 are reinterpreted in the light of those of 1848.

This essay has applied a general political model in place of a professionalization one, partly because contemporaries generally saw events in these terms, and also because these events were distinctly part of political developments which were affecting virtually every quarter of English life. Some trappings of professional behaviour, including the tendency to foster group solidarity, to promote exclusiveness in an elite united regardless of differences in general education, social rank or economic standing, are clearly present. But the agonies of professionalization (*pace* Babbage) – at least in relation to occupations, qualifications and contractual obligations – were not a central factor in English science then, nor indeed, for another four generations. Savants had become more democratically "men of science", but they were still "philosophers", not yet given to the pursuit of what J. A. Thomson later called 'Brodwissenschaft'.[107]

Nevertheless, the institutions of English science did create a route to professional enterprise. This was nowhere clearer than in medicine and in those technical occupations and institutions associated with commerce and the useful arts, where the social recognition afforded philosophers often ensured both profit and status. Many of these professions, however, were similarly dominated by autocratic metropolitan institutions.[108] In these professions, demands for reform were bound to win attention. In the nature of events these demands arose, first, from men who were in sympathy with liberal sentiments and who were determined, for a mixture of reasons, to couple the power of knowledge with commercial manufacturing and trading interests. For this reason, the history of the RS offers a prismatic perspective on those issues of power, status and patronage central to metropolitan culture.

This essay has stressed the particular relation between contemporary reform movements in science and in medicine. Between 1830 and 1850 the London scientific establishment, like the medical establishment, accepted that it must justify its social status by elevation of its philosophical character – in Huxley's words – by raising 'the level of its scientific reputation'. The scientific establishment sustained the new role by asserting the utility of "pure science", and by restricting entry to its elite. In *The Athenaeum*'s phrase, the 'survivors are of the opinion that the smaller the number the greater the honour'.[109] As a result, the RS won increased prestige and government patronage. The medical establishment achieved similar ends by creating a respectable professional image, linked with government sanction, and ultimately justified on grounds of improving standards of science. In medicine, as in science, the major institutions deflected competition and criticism, partly by departmentalizing knowledge in disciplines, and partly by encouraging the universities to defend these departments, and so offer themselves as both social and intellectual selectors for the metropolitan elite.

This essay has also focused on the twin issues of hegemony and patronage. Perkin has described the Victorian era as witnessing the replacement of 'old vertical connections of dependency and patronage' by the 'horizontal solidarities of class'.[110] As we have seen, this process, if it can be said to have ever occurred in science, had scarcely begun by 1850. Vertical connections of class, were, in fact, replaced by vertical connections of family and interest,[111] reinforced not merely by the authority of expertise, but also by restriction of access. The achievement of the RS by 1848 was, in Huxley's words, 'not to create an academy of immortals, but to save the Fellowship of the Society from becoming a strain and an imposition'.[112] In this process, there was an apparent shift from reward by patronage to recognition by merit; from recognition of rank to the reward of ability. But the real significance of this shift should not be exaggerated. While increasing competition and creating a rewards system, the RS did not affect, or seek to affect, the economic base of the would-be philosopher. The "amateur" tradition, preserved by the *corps d'élite* in the RS, naturally persisted as long as that elite remained intact. That elite would eventually be modified in complexion, not principally in economic terms, but in normative expectations. In the process, the RS may be seen to have merely exchanged one form of cultural capital, based on social ascription by class, family and friendship, for another, based on the material and social prerogatives of an educated elite. To complete the transformation, new forms of patronage, operated by representative peers were required. For, as Weld observed in 1848, 'It is the want of patronage that drives science from the halls of our universities; for she has no such rewards to confer on her students as those attached to the bar or to the Church.'[113] Within two decades, the major institutions of science, as other British institutions, had, by accommodating themselves to the threat of rising class interests, successfully set anchors to windward, and had retained for the metropolis, joined in holy trinity with Oxford and Cambridge, primacy both in intellectual and political authority.

It remains to ask why the RS, like the Royal Colleges of Physicians and Surgeons, failed to act until well after 1900, in the interests of the general practitioner or the rank and file of science. By confining its resources to prizes, by confining its elections to virtual prize fellowships, the RS inevitably became part of the competitive struggle for survival in nineteenth-century intellectual life. Yet, in practice, the actual encouragement of specific areas of science was left to others – the learned societies, the BAAS, field clubs, eventually the universities, university colleges and schools. In the debates which filled the pages of *Nature* and supplied the motives of the scientific movement some twenty years later, the RS stayed at a safe distance from "bread and butter" matters. Even a new generation of "Young Turks" in the 1870s and club rule from the famous 'Xs', did not quickly alter the picture. In both cases, it can be argued that neither the medical nor the scientific establishment could have won greater social advantage by advocating greater participation. But in both cases, radical alternatives, seeking a broader base for professional development, were rejected as socially, and therefore, professionally, undesirable. It surprised no one that parliamentary democracy was limited to the confines of a small, closely examined and socially reproduced elite – in Coleridgian terms, a clerisy of science.

Paradoxically, owing largely to the character of its reform between 1830 and 1850, many of the most important developments in late Victorian applied science and science education took place largely outside the compass of the RS. Its successive Councils preferred a policy of *dégagement*, justified by the need for a suitable division of labour, but also by the need to distinguish between industry and insight, between enthusiasm and excellence.[114] As a result, in pursuing its philosophical independence, the RS almost inadvertently underwrote a policy of professional science, increasingly detached from the wider and more popular scientific, technological and educational interests of the country. Science was perhaps not without a head, but it was, as Huxley and others later claimed, without a nervous system. It is ironic that the absence in England of a national system of scientific "organization", lamented for the next fifty years from Matthew Arnold to Arthur Balfour, could scarcely have been more thoughtfully arranged.

Acknowledgements

This paper is a much shortened version of a longer essay which examines more closely the institutional history of the Royal Society and scientific London in the first half of the nineteenth century. For their help and comments the author wishes to thank the editors, Dr Kay Andrews, and Mr N. H. Robinson. For permission to use their manuscripts the author is grateful to Trinity College, Cambridge, the Royal Society of London, and the Royal Institution. Since this paper was submitted, several relevant works have appeared which I have been unable to take into account. These include David Miller, 'The Royal Society of London, 1800–1835: a study in the cultural politics of scientific organisation', unpublished Ph.D thesis, University of Pennsylvania, 1981, and Jack Morrell and Arnold Thackray, *Gentlemen of Science: Early Years of the British Association for the Advancement of Science*, Oxford, 1981.

Notes and references

1 H. Lyons, *The Royal Society, 1660–1940: A History of its Administration, under its Charters*, Cambridge, 1944; D. Stimson, *Scientists and Amateurs: A History of the Royal Society*, New York, 1948.

2 J. T. Merz, *A History of European Thought in the Nineteenth-Century*, 4 vols., Edinburgh and London, 1904, **1**, pp. 230–48.

3 N. Reingold, 'Babbage and Moll on the state of science in Great Britain', *British Journal for the History of Science*, 1968, **4**, 58–64.

4 G. Foote, 'The place of science in the British Reform Movement, 1830–1850', *Isis*, 1951, **42**, 192–208.

5 C. Babbage, *Reflections on the Decline of Science in England*, London, 1830; M. Moseley, *Irascible Genius: The Life of Charles Babbage, Inventor*, London, 1964. For criticisms of the universities: C. Lyell, 'State of the universities', *Quarterly Review*, 1827, **36**, 216–68; and M. Sanderson, *The Universities in the Nineteenth-Century*, London, 1975, pp. 26–72.

6 G. Foote, op. cit. (4), p. 208; also, Foote, 'Science in early nineteenth-century England', *Osiris*, 1954, **11**, 438–54. cf. R. MacLeod, 'Introduction. On the advancement of science', in R. MacLeod and P. Collins, *The Parliament of Science*, London, 1981, pp. 17–42.

7 L. P. Williams, 'The Royal Society and the founding of the British Association for the Advancement of Science', *Notes and Records of the Royal Society*, 1961, **16**, 221–33.

8 E. Mendelsohn, 'The emergence of science as a profession in nineteenth-century Europe', in K. Hill (ed.), *The management of scientists*, Boston, 1964, pp. 3–48. A tendency to ascribe professional attributes to science without a close reading of recent history was commonplace by the 1870s: T. H. S. Escott, *England: Its People, Polity and Pursuits*, London, 1879, p. 452; and H. Spencer, *The Principles of Sociology*, London, 1893.

9 cf. D. Knight, *The Nature of Science*, London, 1976, pp. 92 *et seq.*; and S. F. Cannon's corrective *Science in Culture*, New York, 1978, pp. 137–65.

10 The best review of professionalization is devoted to American

science but applies to Britain: N. Reingold, 'Definitions and speculations: the professionalisation of science in America in the nineteenth-
century', in A. Oleson and S. Brown (eds.), *The Pursuit of
Knowledge in the Early American Republic*, Baltimore, Md, 1976,
pp. 33–69. For the difficulty of assigning professional labels to
different disciplines in Britain, see R. Porter, 'Gentlemen and
geology: the emergence of a scientific career, 1660–1920', *The
Historical Journal*, 1978, **21**, 809–36.

11   I. Inkster, 'Science and society in the metropolis: a preliminary
examination of the social and institutional context of the Askesian
Society of London, 1796–1807', *Annals of Science*, 1977, **34**, 1–32; A.
D. Orange, *Philosophers and Provincials: The Yorkshire Philosophical Society from 1822 to 1844*, York, 1973.

12   W. F. Cannon, 'History in depth: the early Victorian period', *History
of Science*, 1964, **3**, 20–38.

13   J. B. Morrell, 'Individualism and the structure of British science in
1830', *Historical Studies in the Physical Sciences*, 1971, **3**, 183–204.

14   The concept of hegemony, taken from Gramsci's theory of cultural
supremacy, embraces the distinction between the social roles of
"traditional" intellectuals (scholars) and "organic" intellectuals
(including technical and bureaucratic experts): G. Williams, 'The
concept of "Egemonia" in the thought of Antonio Gramsci: some
notes on interpretation', *Journal of the History of Ideas*, 1960, **21**,
586–99; M. Berman, 'Hegemony and the amateur tradition in British
science', *Journal of Social History*, 1975, **8**, 30–50.

15   W. Bagehot, *The English Constitution*, London, 1867, pp. 252–67.

16   S. de Morgan, *Memoir of Augustus de Morgan*, London, 1882, p. 42.

17   "Philosopher" and "philosophical" are used throughout in their
contextual sense to describe those holding a "philosophical" (as
distinguished from a received) view of nature, or a member of a
"philosophical circle".

18   A. Briggs, *The Age of Improvement, 1783–1867*, London, 1959, pp.
184–235; J. Droz, *Europe between Revolutions, 1815–48*, London,
1967.

19   W. L. Burn, *The Age of Equipoise: A Study of the Mid-Victorian
Generation*, London, 1964, pp. 15–54.

20   Bagehot, op. cit. (15).

21   J. Barrow, *Sketches of the Royal Society and the Royal Society Club*,
London, 1849, pp. 16–52, describes Banks's court.

22   C. Lyell, 'Scientific institutions', *Quarterly Review*, 1826, **34**, 153–79;
R. Schofield, *The Lunar Society of Birmingham: A Social History of*

Provincial Science and Industry in Eighteenth-Century England, Oxford, 1963; A. Thackray, 'Natural knowledge in cultural context: the Manchester model', *American Historical Review*, 1974, **79**, 672–709; S. Shapin, 'The Pottery Philosophical Society, 1819–35: an examination of the cultural uses of provincial science', *Science Studies*, 1972, **2**, 311–36.

23   Following the secession from the RS of the Linnean Society in 1788, the disrobing of the "old lady" continued with the creation of the Geological Society (1807), the Astronomical (1820), the Zoological (1828), and the Geographical (1830) Societies.

24   M. Berman, *Social Change and Scientific Organisation: The Royal Institution, 1799–1844*, London, 1978. The interpenetration of the RI and the RS can be deduced from biographical notes included as appendices in Berman's thesis of the same title (Ph.D thesis, Johns Hopkins University, 1971), but not reproduced in his book.

25   The Society of Arts from the 1770s had cultivated good relations with the RS: D. Hudson and E. W. Luckhurst, *The Royal Society of Arts, 1754–1954*, London, 1954.

26   Slightly more physicians (thirty-three in all) were elected to the RS in the decade 1810–19 than in the four previous or subsequent decades: *Record of the Royal Society*, London, 1940; W. Munk, *Roll of the Royal College of Physicians of London*, London, 1861.

27   A. C. C. Swinton, *Autobiographical and Other Writings*, London, 1930, p. 62; *Natural Science*, 1894, **4**, 2–3; A. T. Basset (ed.), *A Victorian Vintage*, London, 1930, p. 124; A. B. Granville, *Autobiography*, London, 1874, **i**, p. 66.

28   Babbage to Whewell, 15 May 1820, Whewell Papers, Trinity College, Cambridge.

29   J. F. W. Herschel, *A Preliminary Discourse on the Study of Natural Philosophy*, London, 1830; S. F. Cannon, op. cit. (9), 29–71.

30   Gilbert (1767–1839) became an FRS in 1791 and soon entered politics. During Fox's administration he attracted praise for his knowledge of public finance and government administration, and for his assistance with inquiries into weights and measures, copyright and mining. During Liverpool's administration, he nurtured good relations between the RS and the Admiralty; through his intercession, more scientific places were created on the Board of Longitude and Babbage received the first grant given for his calculating engine.

31   A. C. Todd, *Beyond the Blaze: A Biography of Davies Gilbert*, Truro, 1967; and 'The life of Davies Gilbert, 1767–1839', unpublished Ph.D thesis, University of London, 1958, p. 654.

32 R. MacLeod, 'Of medals and men: a reward system in Victorian science, 1826–1914' *Notes and Records of the Royal Society*, 1971, **26**, 81–105.

33 Todd, op. cit. (31, 'Life'), p. 698; Todd, op. cit. (31, Biography), pp. 205, 237 (quotation).

34 Herschel to Babbage, 15 December 1829, Herschel Papers, RS, 2, 242.

35 *The Lancet* (1829), 20 June 1829, **1**, 383–4.

36 e.g. "Argus" (?Babbage), *The Times*, 30 June 1830.

37 *The Lancet*, (1829–30), 3 April 1830, **1**, 16.

38 Following the recent death of George IV, the Duke of Sussex was brother to the new King, William IV. Educated at Göttingen, his liberal and cosmopolitan credentials were impeccable. From 1811 as Grand Master of the Freemasons and from 1816 as President of the Royal Society of Arts, he had drawn a closer circle between the Crown and progressive policy.

39 W. Fitton, *A Statement of Circumstances Connected with the Late Election for the Presidency of the Royal Society*, London, 1831.

40 P. Hollis, *The Pauper Press: A Study in Working Class Radicalism*, Oxford, 1970, pp. vii, 30.

41 e.g. G. P. Scrope, Liberal MP for Stroud, 1833–68; and R. Vyvyan, Tory MP successively for Cornwall, Okehampton, Bristol and Helston, 1825–57.

42 Including E. R. Daniell, barrister and Governor of the RI, 1816–31; H. Hallam, historian and barrister; and R. H. Solly, one of the original promoters of the RI.

43 Fitton, op. cit. (39), p. 29. A close examination of the proposed Councils reveals remarkable similarities. Both Sussex and Herschel, for example, would have dropped Sir John Franklin, Sir Everard Home, Sir Thomas Lawrence and Adam Sedgwick; Herschel and Sussex would have both retained Roget as Secretary. They were also agreed to keep George Rennie, John Pond, Henry Kater, Faraday, and Gilbert himself. Both parties would have *added* William Cavendish and John Lubbock. In fact, the area of disagreement centred upon only twelve places. Clearly few new men were involved on either side. Herschel's cabinet was highly philosophical but hardly professional. The Duke's was decidedly representative of the *patronat*. Clearly there would have been differences in spirit but scarcely a revolution in values.

44 A. B. Granville, *Science Without a Head; or, the Royal Society Dissected*, London, 1830.

45 W. Tooke wrote to Sussex asking him to withdraw: Williams, op. cit. (7), p. 230.

46 ibid, p. 230.

47 *The Times*, 1 December 1830, and Gilbert quoted in Todd, op. cit. (31, *Biography*), p. 265.

48 A. B. Granville, *The Royal Society in the Nineteenth Century*, London, 1836, p. 218.

49 Moseley, op. cit. (5), pp. 122–5.

50 Briggs, op. cit. (18), p. 207. P. D. Lowe, 'Locals and cosmopolitans: a model for the social organisation of provincial science in the nineteenth-century', unpublished MPhil. thesis, University of Sussex, 1978, pp. 16 and *passim*. The creation of new provincial societies, a movement beginning in the 1810s and 1820s, reached its first climax in the 1830s.

51 N. Gash, *Politics in the age of Peel*, 2nd edn., London, 1977, p. 10.

52 *Gentleman's Magazine*, 3 May 1834, **1** (n.s.) 540.

53 G. B. A. M. Finlayson, *England in the Eighteen-Thirties*, London, 1969, pp. 31–6.

54 M. J. Durey, 'British Society and the cholera', Ph.D thesis, University of York, 1975, pp. 1–11.

55 E. M. Little, *History of the British Medical Association, 1832–1932*, London, 1932, pp. 117 *et seq.*, and C. Brook, *Battling Surgeon*, Glasgow, 1945.

56 Kater remained a Councillor, but resigned as Treasurer. His successor in 1830, Lubbock, endowed the Society's financial affairs with all the respectability of a successful banker. Barrow provided a useful naval link; Murray, a link with the Tories; Vigors, a link with the liberal landed Whigs; John Rennie, with the engineers; Lord Melville, with the Admiralty and the House of Lords; and Peel, whose parliamentary influence guaranteed a steadying hand, with the forces of moderate reform.

57 Return of Civil List Pensions, 1840, p. 302. His object, Peel explained, to Mrs Somerville, was 'a public one, to encourage others, and to prove that great scientific attainments are recognised among public claims': M. Somerville, *Personal Recollections from Early Life to Old Age, of Mary Somerville*, London, 1874, p. 177.

58 Melbourne awarded no Civil List pensions to men of science during 1837–40. For the insult accorded Faraday by Melbourne: L. P. Williams, *Michael Faraday*, London, 1965, p. 353; and E. P. Hood, *The Peerage of Poverty*, London, 1859, pp. 213–15.

59 Barrow, op. cit. (21), p. 124.

60 J. Douglas, *The Prospects of Britain*, Edinburgh, 1831, p. 59.
61 Between 1833 and 1849, the BAAS assisted over fifty men and 300 projects with sums amounting to £15,000: R. MacLeod, 'The Royal Society and the Government grants: notes on the administration of scientific research, 1849–1914', *The Historical Journal*, 1971, **14**, 323–58.
62 For the political principle of association, in opposition to the autocracy of science: T. Robinson, *Presidential Address to the British Association at Birmingham*, 1849, pp. xxx–xxxi; J. M. Baernreither, *English Associations of Working Men*, London, 1891; F. H. Giddings, *The Principles of Sociology: An Analysis of the Phenomena of Association and of Social Organisation*, London, 1899; E. C. Black, *The Associations: British Extra-Parliamentary Political Organisation*, Cambridge, Mass., 1963; R. MacLeod, op. cit. (6), pp. 20, 38.
63 "FRS", *Thoughts on the Degradation of Science in England*, London, 1847, p. 49.
64 K. M. Lyell, *Life, Letters and Journals of Sir Charles Lyell*, London, 1881, **ii**, p. 45.
65 C. Babbage, *On the Economy of Machinery and Manufactures*, London, 1832, p. 313.
66 Tyndall to Spottiswoode, 21 January 1878, Tyndall MS (RI), II, fo. 1273.
67 Leibig to Berzelius, 26 November 1837, quoted in T. E. Thorpe, *Essays in Historical Chemistry*, London, 1923, p. 588.
68 E. B. Lytton, *England and the English*, London, 1833, p. 17.
69 Granville, op. cit. (27), **2**, p. 224; Peel to Granville, 31 October 1838.
70 Barrow, op. cit. (21), p. 124.
71 Spencer Joshua Alwyne Compton, second Marquis Northampton (1790–1851), was a 'well-educated nobleman, a traveller . . . versed in the laws and institutions of his country . . . fit for any situation that an English gentleman could be qualified to hold': Barrow, op. cit. (21), p. 128.
72 A. Whitridge, *Men in Crisis: The Revolutions of 1848*, New York, 1949; J. Sigmann, *1848: The Romantic and Democratic Revolutions in Europe*, London, 1973.
73 Briggs, op. cit. (18), p. 331; R. Owen, *Life of Richard Owen*, London, 1894, **i**, pp. 230, 236. Murchison was knighted in 1846; Richard Owen was offered, and declined, a similar honour. Peel's scientific weekends at Drayton Manor in the early 1840s were attended by Playfair, Wheatstone, Owen, Brodie and George Stephenson among others: Parker, *Sir Robert Peel*, London, 1899, **ii**,

304–7; **iii**, pp. 162, 225, 433–5, 447. In the 1840s De la Beche, Playfair, Owen, Horner and others repeatedly gave evidence to select committees on questions ranging from smoke and noxious vapours to factory and health legislation: W. M. Fraser, *A History of English Public Health, 1834–1939*, London, 1950, p. 19. Peel also used William Buckland as his informal scientific adviser. Babbage, alas, was disappointed.

74   A. de Morgan, 'Science and rank', *Dublin Review*, 1842, **13**, 275–6.

75   *Sharpe's London Magazine*, 1845, **280**, p. 581.

76   Lyons, op. cit. (1), pp. 257–8.

77   Granville, op. cit. (27), **2**, p. 222.

78   Barrow, op. cit. (21), pp. 127, 132.

79   Northampton to Herschel, 26 May 1845, Herschel Papers, RS, HS 5. 266.

80   The correspondence was published, with editorial comment supporting Lee against the Society in *The Lancet*, 1846, **1**, 526–9, 583–5.

81   *The Lancet*, 1846, **2**, 408.

82   The committee was chaired by Northampton and included the Vice Presidents (Rennie, Lubbock, Wrottesley and Hooker), the two Secretaries (Roget and Christie) and the Foreign Secretary (Sabine) as well as Grove himself.

83   *The Lancet*, 1846, **2**, 408; Lyons, op. cit. (1), p. 260.

84   Z. Cope, *The Royal College of Surgeons of England: A History*, London, 1959, pp. 64–7.

85   D. Martin, 'The Royal Society today', *Discovery*, 1960, pp. 292–3.

86   L. Levi, 'On the progress of learned societies, illustrative of the advancement of science in the United Kingdom during the last thirty years', *Report of the British Association for the Advancement of Science*, 1868, 169–97; *Daily News*, 27 April 1871. In that year there were fifty candidates, of whom nineteen were physicians, five chemists, four geologists and one a mathematician.

87   Granville, op. cit. (27), **i**, p. 66.

88   The Council of 1846–7 still included Brande, of Davy's day, and George Rennie, Roget and Paris, of Gilbert's Councils; but it also included a coalition of philosophers, e.g. Samuel Christie, Samuel Cooper, Edward Sabine, Henry De la Beche, Edward Forbes, William Hopkins, George Porter, Baden Powell, Sir John Richardson, W. H. Sykes and W. H. Smyth. Four men (Cooper, Forbes, Porter and Richardson) were on the Council for the first time.

89   J. P. Gassiot to Herschel, 5 March 1847 and 1 April 1847, Herschel Papers, RS, HS 8. 55–6; T. G. Bonney, *Annals of the Philosophical*

Club of the Royal Society, London, 1919; T. E. Allibone, *The Royal Society and its Dining Clubs*, Oxford, 1976, pp. 199–205.

90   The "forty-seven" included five surgeons and two physicians; another six were then professors at King's College, London; four others were professors at UCL; and four were governors of the RI. Among ten geologists or palaeontologists, three (Horner, Lyell and De la Beche) were, or would become, Presidents of the Geological Society. Remarkably, none of the officers of the RAS or the Linnean was represented. The Club insisted upon an accommodating policy towards the learned societies. Forbes urged Grove to avoid setting dates of meetings to clash with other societies: it was 'very important that they should gradually learn to look on the Philosophers as a sort of higher council or guardian angel for them all': Forbes to Grove (n.d., probably 1849), Grove Papers, RI.

91   In 1845–6 the twenty-one officers of the RS included eight members of the new Club; in 1846–7 there were fourteen, including the Treasurer (Rennie), the Junior Secretary (Christie) and the Foreign Secretary (Sabine). Of the forty-seven, only seventeen had been Fellows in 1830, and not all were administrative reformers; only five had signed the Fellows' Requisition, and only nine signed Herschel's Declaration that year. Of the forty-seven, thirty had served on the RS Council between 1831 and 1846. By their second meeting on 6 May, thirteen additional Fellows had accepted, and by June their numbers were complete.

92   *The Athenaeum*, 27 May 1848, p. 509.

93   W. White, *The Journal of Walter White*, London, 1899, p. 82.

94   *Punch*, 18 March 1848.

95   Northampton to Rosse, 21 March 1848, Herschel Papers, RS, HS 5. 274. According to Lyell, Faraday declined to stand, but was prepared to support either Robert Brown or Richard Owen as 'good men of European reputation': Sabine to Grove, 1 February 1848, Grove Papers, RI; "FRS", op. cit. (63), p. 59; Rosse to Sabine, 6 November 1854, RS, Mc. 5. 181; Rosse to Sabine, from Birr Castle, Parsonstown, 23 March 1848, Sabine Papers, RS, SA 1112. On Sabine, see N. Reingold, *Dictionary of Scientific Biography*, **12**, 49–53.

96   Sabine to Grove, 28 November 1848, Grove Papers, RI.

97   *The Athenaeum*, 27 May 1848, p. 509.

98   Carpenter to Grove, 22 November 1848, Grove Papers, RI.

99   White, op. cit. (93), p. 86.

100   Minutes of the Government Grant Committee, 11 November 1849, cited in MacLeod, op. cit. (61). The model for the parliamentary

grant was the voluntary research fund begun over a decade before by the BAAS: Minutes of Government Grant Committee, 7 March 1850. The vocabulary of voluntarism had been assimilated into the grammar of government.

101   The appointment of assistant secretaries of the calibre of Charles Weld (1843–61), and Walter White (1861–85), the publication of accounts, and the expectation of regular attendance from the two Secretaries and the Foreign Secretary in return for substantial salaries, all sustained an image of greater diligence: A. Strange, 'The government of the Royal Society', *Nature*, 1870, **3**, 1–2; White, op. cit. (93), p. 209.

102   A. Hume, *The Learned Societies and Printing Clubs of the United Kingdom*, London, 1853, pp. 41–2.

103   D. C. Moore, 'Concession and cure: the sociological premises of the first Reform Act', *Historical Journal*, 1966, **9**, 39–59; and his *Politics of Deference*, London, 1975.

104   *The Times*, 29 November 1848.

105   Cannon, op. cit. (9), p. 145.

106   Berman, op. cit. (24). For alternatives worth pursuing, Cannon, ibid., pp. 150–63.

107   J. A. Thomson, *Progress of Science in the Century*, London, 1908, p. 46.

108   cf. R. N. Shaw and T. G. Jackson (eds.), *Architecture: A Profession or an Art?*, London, 1892; P. Smith, *History of Education for the English Bar*, London, 1860; F. C. Thompson, *Chartered Surveyors: The Growth of a Profession*, London, 1968; B. Heeney, *A Different Kind of Gentleman: Parish Clergy as Professional Men in Early and Mid-Victorian England*, Hamden, Conn., 1977.

109   *The Athenaeum*, 23 December 1865, p. 891.

110   H. Perkin, *Origins of Modern English Society, 1780–1880*, London, 1969, p. x.

111   As late as 1874, Galton found that 120 of 180 FRSs then alive belonged to only thirteen different families: *English Men of Science*, London, 1879, p. 40; and his *Index to Kinsmen of FRSs*, London, 1904.

112   T. H. Huxley, *Proceedings of the Royal Society*, 1885, **39**, 281.

113   C. R. Weld, *A History of the Royal Society*, London, 1848, **ii**, p. 467.

114   RS Council Minutes, 30 November 1875, Report of Election Statutes Committee.

3 The London lecturing empire, 1800–50

J. N. Hays

In the first half of the nineteenth century London's dominance of the scientific life of Great Britain increased. This dominance was founded on London's place as the home of new national organs and organizations of science, scientific culture, and scientific diffusion. By 1850 a large number of societies had been created, national in scope but London-based, which catered to the growing specialization of science and to the nascent professionalism of those who practised it. Periodicals edited and published in London increasingly dominated the diffusion of scientific ideas to the literate of the country. Societies based in London undertook the publication and distribution of tracts and series of volumes informing readers of science. London publishing houses contributed to the same end. Government was hesitatingly extending its arm into the scientific community. Institutions which in one way or another were regarded as national repositories – whether of wisdom, wondrous technology, or wild animals – abounded in London. The growth of science in the provinces, connected there with both rapid industrialization and the scientific reform movement, has deservedly claimed the attention of modern scholars.[1] But the centripetal pressures toward the metropolis were powerful indeed, and for that reason more attention should be given to the workings of the scientific community within London.

For the non-gentlemanly members of the scientific communities of London between 1800 and 1850 three activities were especially important: scientific lecturing, medical practice, and technological involvement in trade and industry. I shall argue here that scientific lecturing in London was decisively institutionalized by the 1820s; that institutionalization of lectures contributed to the support and hence the professionalization of men of science; that institutionalization also entailed the payment of a price for that support; and that institutionalization brought London lecturing more clearly within the influence of the other important components of London science, namely, medicine and industrial technology. I shall further argue that London popular science, partly because of its institutionalization and partly because of London's position as the centre of spectacle, was able to offer to provincial Englishmen views of science which they could find nowhere else.

I

In what ways did non-affluent men of science in London find support for their activities? Few men of science in London (or indeed any place in Britain) enjoyed a salaried position as such. The Government supported the Astronomer Royal and his assistants, and military men engaged in scientific enterprises such as the Ordnance Survey; from the 1830s it offered Civil List pensions to some men of science.[2] The Royal Institution, from its foundation in 1799, sustained a few others.[3] The professions supported others, and the fact that London dominated the nation's professional life brought some men of science to it. Charles Lyell was a barrister, although he practised little; William Robert Grove was a barrister too, and became a judge.[4] But medicine was far more important. London's place in the nation's medical world was paramount; membership in the London-based Colleges of Physicians and of Surgeons was the social pinnacle of the profession, and London hospitals supplied an increasingly important and organized body of medical teaching for the entire country.[5] Medical practice and teaching supported a large fraction of the London scientific community – and not all of them involved in the biological sciences either, as can be seen in the diverse careers of Thomas Young, William Hyde Wollaston, and George Birkbeck.[6]

Although it is true that much of the pioneering technology of industrialization was provincial and not metropolitan, the diverse structure of London industry and trade in fact afforded considerable support for participants in its scientific life. Sometimes trade simply afforded an income that made a scientific avocation possible, as stockbroking did for the astronomer Francis Baily.[7] But London industry and trade had more direct associations with science. Just as medicine could be both source of income and a science-related activity in itself, so trade or industry could serve both purposes. London had long been the home of precision industries, of shipping (both associated with astronomy), and of the chemical trades: William Allen was both manufacturing chemist and natural philosophy lecturer. In the early nineteenth century such new industries as machine tools and gas lighting were London-centred. The place of technology in London life, and of London technology in the nation's industrialization, should not be underestimated.[8] And because the demarcation lines between "science" and "technology" were not especially clear, much of this type of activity was regarded as "scientific".

For many men of science teaching of some sort was an important aspect of London's appeal. This teaching was conducted on many levels. We certainly do not have to wait for the foundation of University College (1826) and King's College (1829) to find scientific teaching in London.[9] Although in the words of a modern authority, 'Teaching hospitals, in the modern sense of the word, did not exist in 1800', the same author describes eighteenth-century lecturing at St Thomas', St Bartholomew's, Guy's, London, and Middlesex Hospitals by that date.[10] Between 1800 and 1840 all these hospitals developed formal medical schools, an endeavour in which they were joined by St George's, Westminster, Charing Cross, University College, and King's College Hospitals. This system of medical education employed the talents of physicians, surgeons, apothecaries, chemists, and botanists, both in the hospital schools and on an independent freelance basis. In addition to formal medical education, there existed in London a growing number of institutions which offered scientific lectures to their members, and which paid men of science to deliver the lectures.[11] These institutions varied from the private lecturer who performed in rented rooms, which he dignified with an institutional name, to the large institution which paid a salaried employee to deliver lectures to a fee-paying membership. This dense network of lecturing in the metropolis in fact constituted a major source of support for members of the scientific communities within it. Still another form of teaching, writing for a wide public, had its centre in London.

London provided the sustenance of intellectual community for men of science, as well as mere food for their bodies. A growing collection of London-based, but national, societies catered to the increasingly specialized divisions (and even subdivisions) of the sciences: the Linnean Society (1788), the Geological Society (1807), the Astronomical Society (1820), the Zoological Society (1826), the Chemical Society (1841).[12] Their meetings were a magnet for the scientifically inclined. London social life included what might be styled scientific attractions, some profound and some simply silly, where the vague line between scientist (itself a word coined only in 1833)[13] and laymen might vanish altogether, or where the scientist could show off his wares to an admiring laity: the Thames Tunnel (1828), balloon ascents, Professor Wheatstone's telegraphic device at King's College, Professor Owen's bones at the College of Surgeons, Herschel's great telescope at Slough, Babbage's calculating machines, a whale stranded in the Thames, or the corpse of Jeremy Bentham, publicly dissected by Southwood Smith.[14] A number of men of science (and at least one woman) formed part of a more general London elite, members of a Coleridgean clerisy, the Tight Little World of the early and middle nineteenth-century learned and leisured classes. Nowhere but in London was such social interchange possible for the man of science. London was also the place of opportunity where a name could be made or lost. Some men of obscure origin – Humphry Davy, Michael Faraday, Richard Owen, for instance – triumphed in London. Others failed: John Millington encountered financial trouble and went to Mexico; John Frost was called a charlatan and went to Berlin to pose as a physician; Dionysius Lardner, a great success in the 1830s, became involved in a disastrous love affair; John Gordon Smith's career as a lecturer failed, he turned to drink and died in a debtors' prison. London was the last resort of the desperate and the scorned: Robert Knox, the unfortunate Edinburgh anatomist who purchased specimens from Burke and Hare, attempted to recoup his fortunes by lectures and practice in London.[15] London, in short, was the great city, alluring alike to social lions and to misfits.

London scientific life was dominated above all by the lecture. The number and types of scientific lecture in London changed considerably in the first half of the century. Much of this change was accomplished in the 1820s and 1830s, by which time a chaotic welter of lectures was being replaced by a more highly structured and institutionalized world. Inkster's careful itinerary of London lectures at the turn of the century demonstrates an impressive number and variety of scientific lectures, most of them conducted under private auspices in homes or rented halls, in different portions of London.[16] There were relatively few formal institutions at which lectures were performed: the Royal Institution (1799), the Russell Institution (1808), the Surrey Institution (1810), the London Institution (1805, but no lectures before 1819), plus literary and philosophical societies in Hackney and Marylebone. Other organizations were more or less formal, but retained something of the character of private individual ventures: Samuel Varley's chemical demonstrations (1794) became the London Philosophical Society, and then the Philosophical Society of London (1811); the City Philosophical Society (1809), at the centre of which was John Tatum, whose house was open to members; the London Chemical Society (1806), the forum of Frederick Accum and later of Frederick Joyce; the Askesian Society and its near-relative, the British Mineralogical Society, both somewhat informally organized. A number of private lecturers presided over institutions that were personal ventures; others plied their trade in their own homes, or in rented premises, as well as lecturing to more formal organizations.

The subjects of these lectures may be divided into several categories, which in practice overlap. Much of the lecturing activity grew out of London's position in the English medical world. In the first decade of the nineteenth century the most elaborate programme of lectures was offered by the adjoining St Thomas' and Guy's Hospitals, where a number of men of science lectured within an integrated curriculum.[17] Other hospitals sponsored lectures as well – London, Middlesex, St George's, St Bartholomew's. "Theatres of Anatomy" in Great Marlborough Street and in Great Windmill Street (the latter originally Hunter's private medical school) employed other lecturers, while individual physicians and surgeons such as George Pearson and Thomas Garnett joined in medical education by conducting lectures on their own premises.[18] This plethora of lecturers in fact served a broader educational function than the mere professional training of physicians and surgeons. Mr Blair advertised his lectures on anatomy and animal economy as being suitable for 'scientific persons, amateurs of natural history, students in the liberal arts, and professional men in general'.[19] The medical lecturers did not restrict themselves to anatomy, physiology and the practice of medicine and surgery. Chemistry and botany were staple items as well, and the world of chemical lectures for medical students overlapped those of other chemical interests; Inkster has connected the spheres of chemistry, electricity and mineralogy in the lectures of such men as William Allen, the Aikins, Richard Phillips, and William H. Pepys.[20] Garnett lectured on both medicine and natural philosophy.[21] The medical possibilities of electricity were a frequent theme – perhaps symbolic of that association were Giovanni Aldini's lectures on galvanic electricity given at St Thomas' Hospital in 1802–3.[22] In a letter to the *Philosophical Magazine* in 1802, Richard Teed (a Strand jeweller) discussed the relief of lumbago by a galvanic belt – illustrating that the connections between medicine and electricity were appreciated by a circle greater than that of the medical practitioners.[23] Chemistry and mineralogy provided a further link between the medical character of London lectures and their utilitarian character.

II

By the late 1820s the picture of scientific lecturing in London had changed most notably in the direction of formalization. It is true that there still flourished a large number of private scientific instructors and entertainers. But overshadowing private enterprise was collegiate and institutional activity. A scientifically minded person, such as the London solicitor Daniel Moore (1759–1828), could find his 'chief amusement' in learned societies, 'where his good humour and love of science always insured a hearty welcome'.[24] Such a person had a rich menu placed before him. The menu was largely seasonal in character, as was London social life; scientific societies and institutions concentrated their activities between November and June. Mechanics' Institutes were not so bound, but other organizations were, even if their membership had no connection whatever with the gentle County-Mayfair society whose peregrinations resulted in the Season and for whom the Season had a rationale.[25] During that Season, by the late 1820s, there were generally three scientific societies meeting per week: the Royal Society weekly at Somerset House, the Linnean Society and the Zoological Club alternating at the former's quarters in Soho Square, the Horticultural Society every other week in Regent Street, the Geological Society alternate Friday evenings in Bedford Street (in Somerset House by the 1828–9 Season), the Astronomical Society monthly in Lincoln's Inn Fields, the Medico-Botanical Society in Sackville Street.[26]

Learned practitioners and their patrons had, therefore, formalized their associations. The arrangements for science of a more popular or educational character had become formalized as well. The metropolis offered a wide range of scientific lectures that afforded institutional instruction and entertainment. The Royal

Institution, firmly based in Albemarle Street, presented between six and ten separate lecture courses through the Season, and had recently instituted Friday evening lectures that quickly became highlights of London scientific-social life.[27] The London Institution, the City's answer to Mayfair's Royal Institution, had moved into an imposing building in Moorfields in 1819, and it too offered a number of lecture courses throughout the Season, as well as Friday evening *conversazioni* on the Royal Institution model.[28] The Russell Institution in Bloomsbury and the Surrey Institution near Blackfriars Bridge were modest versions of these elaborate establishments which also offered scientific lectures; so too did the ancient Gresham Institution (as it was then called) in the City, which had an erratic history in the eighteenth century but which still presented occasional lectures on scientific subjects to the general public.[29] The metropolis had also acquired its own university by the late 1820s, the Godless institution in Gower Street that became University College; scientific instruction was a self-consciously important aspect of that foundation, as it was for King's College, the Anglican counter-blast to the non-sectarian trumpet.

London had also been the centre of vital developments in the movement for Mechanics' Institutes, developments which had provincial reverberations. The London Mechanics' Institution had begun its extensive lecturing, heavily scientific, in 1824 at its quarters in Southampton Buildings, Chancery Lane.[30] As an exemplary model this Institution exerted national influence. Edward Baines is said to have promoted the Leeds Mechanics' Institution under the direct inspiration of George Birkbeck's London lectures; and the aged Birkbeck was greeted as a patron saint by the Manchester Mechanics' Institution in 1840.[31] Within the metropolis itself other Mechanics' Institutes quickly arose in the 1820s: Bethnal Green, Camberwell, Deptford, Hackney, Hammersmith, Poplar, Rotherhithe, Southwark, the City, Spitalfields, Stepney and Westminster all had Mechanics' Institutions or "literary and scientific institutions" active by 1830.[32]

The host of lectures offered to medical students still existed on the London scientific scene in the late 1820s, but they too had a more institutionalized appearance. St Bartholomew's, St Thomas' and Guy's, and Middlesex Hospitals offered full courses of lectures.[33] The Royal Institution had become the venue for some of the lectures previously offered at the Theatre of Anatomy in Great Windmill Street; a variety of other 'theatres of anatomy and medicine' existed as well, on both sides of the Thames, concentrated in Soho and around the large hospitals which themselves offered instruction.

It was not only lectures that were institutionalized in the 1820s and 1830s. Other aspects of the presentation of science to the public moved in the same direction. Private menageries gave way to the Zoological Society of London (1826); private cabinets of natural history gradually gave way to the collections of the British Museum; the Adelaide Gallery (1832) and the Polytechnic Institution (1838) supplanted smaller collections of technological gadgetry.[34] The Society for the Diffusion of Useful Knowledge (1826) attempted to systematize the publication of scientific tracts for the masses.[35] Lardner's 'Cabinet Cyclopedia', begun in 1831, was another massive publishing venture which hoped to systematize scientific information.[36]

Many of these developments confer a watershed character on the London science of the 1820s. The decline-of-science agitation of the 1830s, and the attention paid to it by historians, has perhaps masked the importance of events in the prior decade.[37] For it was in the 1820s that the foundation of University and King's Colleges, the Society for the Diffusion of Useful Knowledge, and the London Mechanics' Institution, combined with the more general institutionalization of scientific activities, set the pattern to which the London scientific scene adhered for the next several decades. That pattern had consequences for the London scientific community, for the diffusion of scientific ideas, and for the relationships of London and provincial science.

For some members of the London scientific community, institutional lectures meant a relatively secure source of income. Private lectures held perhaps more than the usual degree of risk associated with free enterprise. But institutional lecturing was generally undertaken on a contract basis, with the institution paying a flat per-lecture rate. Institutional lecturing would probably not make a man of science rich, although it is true that Humphry Davy's salary at the Royal Institution reached £500 in the first decade of the century.[38] But that was exceptional; Davy's successor William Thomas Brande was paid £200 until 1824, when his salary fell to £150.[39] Davy, Brande (and Michael Faraday, not well paid either) were full-time employees of the Institution, with other duties in addition to lecturing. (It is also true that the other duties offered other sources of income; Brande used the Institution as a locus for his private chemistry lectures to medical students.)[40] But if the Regency splendour of Davy was not to be repeated in later decades, a larger number of more pedestrian figures were subsidized by institutional lecturing; the transition from the flamboyant Davy to the featureless Brande (or the earnest Faraday, for that matter) was symbolic of income as well as purpose. A few examples of some institutional lecturers will make the point.

John Wallis was an important and popular lecturer on astronomy in the 1820s and 1830s. His fees were relatively modest, but his popularity was very great and institutions were delighted to employ him. His colourful spectacles found particular favour at the London Mechanics' Institution, which paid him twenty-four guineas in 1830 and twenty-seven (for six lectures) in 1832, 1834, 1836 and 1838.[41] His fee to the London Institution was higher – forty guineas for the same six lectures; he lectured to the Royal Institution (in 1826 and 1839), to the Marylebone Literary and Scientific Institution (in 1834), and to the Spitalfields, Hackney, and Southwark Mechanics' Institutions.[42] The managers of the London Institution and London Mechanics' Institution were pleased; on one occasion at the Mechanics' Institution the crowd clamouring to attend was so great that Wallis had to repeat his lecture the following day.[43]

A Scottish divine, William Ritchie (1790–1837) was for seven years a dominant London lecturing personality. He began lecturing on the physical sciences at the Royal Institution in 1830 and was regularly employed there until his death, retained as a professor but paid on a per-lecture basis that brought him between £50 and £80 annually.[44] Similar income came from similar services at the London Institution (1834–7); somewhat less (for fewer lectures) from the Mechanics' Institution (1832–7), which paid him five guineas per lecture while the Royal Institution rate seems to have been between seven and eight guineas.[45] Lectures to these three institutions alone gave Ritchie about £150 per year, to which he added the income from other lectures (to the Marylebone Institution, for instance), as well as from a professorship of natural philosophy at University College.[46]

Ritchie's predecessor as a physical lecturer in the capital was John Millington (1779–1868), whose complex lecturing activities did not prevent him from falling into financial difficulties that forced him to flee to Latin America in 1829. Before that he had lectured annually to the Royal Institution from 1815, earning as much as 200 guineas in 1820, as professor paid on a per-lecture basis; £80 from the

London Institution in 1822 supplemented 100 guineas from the Royal Institution in that year; the London Mechanics' Institution paid him five guineas per lecture for a number of appearances in its early years, and the Surrey Institution also called on his services.[47] He was appointed the first Professor of engineering at University College but did not take up the position, claiming that the salary was inadequate.[48] Much later in life he did take up a professorship at the College of William and Mary in Virginia.[49]

Another omnipresent physical sciences lecturer was Charles Frederick Partington (d. 1857), who in the early years of the London Institution served as its assistant librarian.[50] He also received £5 or five guineas per lecture for short courses there through the 1820s and early 1830s, and in addition the Institution paid him for his regular supervisory attendance at its Friday evening soirées.[51] He lectured to the Mechanics' Institutions of London, Spitalfields, and Hackney; to the City of London Literary and Scientific Institution; and to the Surrey and Russell Institutions as well.[52] The London Institution was also the main base of Edward William Brayley (1802–70), joint editor of *Philosophical Magazine* after 1841, who was eventually appointed the London Institution's librarian.[53] His lectures were both frequent and various; to the Royal, London, Mechanics', Russell and Marylebone Institutions, as well as to the Architectural Society, where in 1839 he lectured on the 'properties and natural history of mineral substances used in building and sculpture'.[54] John Hemming, President of the Marylebone Institution, lectured frequently to the London Institution and to the London, Spitalfields, and Southwark Mechanics' Institutions; the London Mechanics' Institution made repeated payments of four guineas per lecture to him through the 1830s.[55]

To this survey of lecturing incomes might be added many others, including many who held positions in the London world and supplemented that position with lecture income: the physician Peter Mark Roget; Gilbert Burnett, Professor of botany at King's College; Robert Grant, Professor of anatomy and zoology at University College; Dionysius Lardner, Professor of natural philosophy at University College (who took up institutional lecturing, as well as prolific authorship, after leaving the College in 1829); John Lindley, Professor of botany at University College; Jonathan Pereira, apothecary to St Bartholomew's Hospital; James Rennie, Professor of natural history at King's College; Charles Wheatstone, Professor of experimental physics at King's College. With the exception of the medical areas, incomes from professorships at the new colleges were extremely insecure, and institutional lecturing offered an important supplement or even alternative. The institutionalization of lectures in London not only insured the incomes of this subclass of performers, but it increased their opportunities to extend their activities into provincial cities.

<h2 style="text-align:center">III</h2>

But did institutions extract a price for their support? Were lectures (and ideas) forced onto Procrustean institutional beds? To some extent the answers to those questions may have been "yes"; but it was a very qualified "yes". As the stimulating work of Berman, Shapin and Barnes suggests, institutions may have been able to impose an ethos on their listeners.[56] But in some respects most institutions that provided lecturing opportunities to men of science were not in a very strong position to impose anything on their performers. If men of science were to some extent dependent on institutional income, institutions in their turn were partially dependent on performers' whims. Institutional "demands" on lecturers apparently grew more out of pressures – especially financial pressures – on the institutions themselves, than out of ideological conviction or class interest. It is possible to say that institutional courses of lectures tended to become shorter in length and hence more restricted in scope; that as a consequence the sciences were presented in a more specialized light; that certain branches of the sciences were more amenable to such treatment than others; and that London institutions were fortunately able to draw on the special skills of the metropolis for those certain sciences. At the same time the style of scientific lecturing underwent some changes, becoming less fervidly rhetorical and more matter-of-fact; and while the institutionalization of lectures was not primarily responsible for that change, it may have contributed to it.

By the 1830s institutional lecturing was moving away from the rhetorical style perfected earlier in the century by Humphry Davy, who in his scientific lectures appealed to moral elevation, poetic inspiration, mental cultivation, business profit and amusement in his long-winded declarations of purpose.[57] Many traces of Regency grandeur lingered on, especially in the pompous pronouncements about the "March of the Mind" issued in Mechanics' Institutes from such as Dr Birkbeck, a contemporary of Davy.[58] But the rhetorical flourishes of Davy's generation did give way, and in the 1830s the exemplary scientific lecturer was not Davy but Michael Faraday, with his clarity, neatness, arrangement, and concentration on the subject at hand.[59] Thus lectures on chemistry (for instance) now discussed properties of elements and compounds, their applications and modes of preparation, while omitting the beauty, hand of God, and happiness of the human race which for Davy were essential to the subject.

Part of the explanation for the change may be found in the changed circumstances of lecturing itself. In the first decade of the century lecture courses, whether privately offered, under the auspices of some informal "institution", or within a relatively structured body such as the Royal Institution, tended to be lengthy affairs. The Royal Institution's first lecturers, Thomas Garnett and Thomas Young, delivered between fifty and 100 lectures per year on the grand theme of "natural philosophy" – an expansive programme which left much time for bombastic rhetoric and declarations of intent.[60] By the 1820s and 1830s, the rapidly multiplying number of London institutions expanded the opportunities for lectures; established performers such as Millington, Ritchie, or Lardner spread themselves thinly over a number of institutions (both metropolitan and provincial), offering a short, select, more specialized repertoire and simply duplicating from one institution to the next. The course of no more than a dozen lectures could thus be far more lucrative to Millington (for instance) than was the vast compendium of natural knowledge presented by Thomas Young at the Royal Institution in 1802–3. What was advantageous to the lecturer evidently pleased institutional managements as well, for brief courses were cheaper and might be more attractive to an organization's membership.

Shortness of funds was an almost universal institutional complaint in the first half of the nineteenth century. A great range of organizations having to do with science soon discovered that they were under-capitalized from the start. The Royal Institution and the London Institution started as proprietary bodies; the Royal Institution rapidly expended the capital raised from the sale of shares to proprietors, and was forced to become a public body to which members paid annual contributions.[61] The London Institution preserved more of its capital intact, but (perhaps with the Royal Institution's example before it) cautiously avoided eating further into that capital, and it too became a public body.[62] The

London Mechanics' Institution was founded virtually without capital at all; it undertook building improvements with some sizeable donations from men of wealth, the acceptance of which bitterly divided the champions of the working classes; but aside from those donations its annual budgets were massively dependent on the dues paid by the membership.[63] Such diverse bodies as the Athenaeum Club (founded in 1824), University College, and King's College had the same experience – inadequate initial capitalization which was consumed by building costs, and resultant dependence on the continuing financial support of members.[64]

For all these bodies – regardless of their many and profound differences – insufficient initial capital meant reliance on fee-paying members. This was most clear in the case of the London Mechanics' Institution, both because of its great dependence on fees and because of the extremely fluctuating character of its membership. Its members paid quarterly dues, generally 5s. or 6s. per quarter; the membership fluctuated wildly, as 300 or more members could be lost every quarter to be replaced by an approximately equal number of new members. This ordinarily rapid turnover could be exaggerated by economic trends. Thus the Institution had 1254 members in June 1836, and 883 in June 1839; its membership fell from 1347 to 1067 in six months in 1827.[65] Other more middle-class bodies which collected annual fees were more stable, but only in degree. It is clear that for all of them institutional survival was in part a question of presenting programmes that were both inexpensive and attractive to a fickle, potentially expanding (or contracting) audience. It should also be noticed that all these organizations operated on extremely tight budgets which left little reserve for emergencies. When crises arose – when, for example, the London Institution had to reconstruct its gas-lighting system – institutions would respond by mutilating their scientific lecture programmes.[66]

Institutional lectures were thus in part dictated by the necessity to make them popular at minimal cost. A possible solution lay in shorter courses which were for the institution cheaper courses, as lecturers were usually contracted on a per-lecture basis. But even if total costs remained the same, the institution gained variety and hence the possibility of pleasing more palates; five lecture courses of six lectures each obviously constituted a more varied programme than two courses of fifteen lectures each. By the 1830s a course of six lectures was a common phenomenon in institutions of many types, and such courses practically forbade systematic surveys of "natural philosophy" or "chemistry"; they encouraged instead a more minute examination of a subdivided field, such as "optics" or "galvanism".[67] Biological and technological subjects were especially suitable. A small biological subdivision – 'Structure, habits and instincts of insects' – could be presented in a few lectures, without the necessity of tying the subject to a larger whole, and without the necessity of assuming prerequisite mathematical or technical knowledge on the part of an audience which might be casual in both its attendance and its attention. Technological wonders – the steam engine, railways, textile machinery, the safety lamp – had the same advantages; and they too, like biological subjects, might be easily and pleasingly illustrated. The popularity of lectures on astronomy demonstrates the tendency towards more narrowly defined and illustrable subjects. As earlier private astronomical lecturers, such as Adam Walker and R. E. Lloyd, had demonstrated, the subject could be dramatically illustrated by illuminated transparencies; and institutional lecturers of the 1820s and 1830s, such as John Wallis, adopted the techniques within a six-lecture framework.

The forces making for shorter courses of lectures also favoured lectures that were single performances. Much of the Royal Institution's *éclat* as a high-level centre of popular science stemmed not from its courses of lectures, but from its Friday evening *conversazioni* begun in 1825. Their popularity was quickly established; the Institution's management credited them with a major role in attracting enough new subscribers to lift the Institution out of a long-standing debt.[68] *Conversazioni* in the 1830s discussed a very wide range of topics, but new technology was a dominant theme.[69] The London Institution attempted to duplicate the Royal Institution's Friday evening programme with its own Friday evenings, begun in 1828.[70] While the success was not as great, the results for the presentation of science were similar: fragmentation, specialization, concentration on easily illustrable technological or biological subjects.

The Royal and London Institutions were powerful and well-established bodies, at least when set beside the range of other literary and philosophical societies and Mechanics' Institutes that multiplied in London. The latter groups often found themselves with single-lecture performances not from choice, but from force of circumstance. The fluctuations of membership, and the shoestring character of budgets, imposed what Thomas Kelly has happily called a 'Micawberish' approach to the presentation of scientific lectures.[71] The lecture programmes of the London Mechanics' Institution – surely the largest London mechanics' institution, and probably a relatively well-organized one as well – often presented what turned up. It employed six lecturers in its first year, 1824; all six men delivered lengthy courses that treated large areas of the sciences – 'Chemistry', 'Astronomy', 'Principles of mechanical science'.[72] By 1826 the number of lecturers had risen to fifteen, the number of courses to twenty-one – a total of ninety lectures during the year, or between four and five lectures per course.[73] Between 1831 and 1839 a pattern was clearly established – roughly ninety lectures per year, delivered by approximately thirty lecturers, in forty to forty-five "courses"; between two and three lectures per "course".[74] In different ways, circumstances drove institutions both mighty and lowly, in metropolis and provinces, to present scientific lectures in more and more discrete pieces; pieces, furthermore, which lacked coherence.[75]

IV

The pieces which were emphasized did in fact bear some relationship to nineteenth-century London life. That London lecturers should devote attention to the biological sciences and to new technology was consistent with the position of the city in the medical and industrial realms. Institutions might have been led to concentrate on biological subjects simply because of the lecturers available in the London medical community. At the beginning of the century much of the freelance scientific lecturing of London was in fact aimed at medical students. As the first half of the century proceeded, medical education gradually became more organized within the shelters of the major hospitals of the metropolis.[76] If anything the position of London in the world of medical education was strengthened, and London was more than ever the magnet for medical practitioners. Institutions which offered scientific lectures had a large population of physicians, surgeons, and apothecaries to draw on as lecturers, and still other men of science – chemists and botanists, especially – found employment as lecturers both in hospitals and in scientific institutions.

A survey of the principal biology lecturers active in London in the 1830s bears out the medical connection; few fell outside it. The list includes the physicians, Robert Edmond Grant, Peter Mark Roget, Thomas Southwood Smith, Robert Dickson, George Birkbeck and

George Lipscomb; the surgeons Thomas Rymer Jones, Samuel Solly, Anthony Carlisle and William Coulson; the apothecary (and later physician) Jonathan Pereira; a surgeon who was Professor of botany at King's College, Gilbert Burnett; botanist John Lindley, who lectured to the Apothecaries' Company (as well as being Professor at University College and involved in medical education there); botanist Charles Johnson, lecturer at Guy's Hospital; and botanist John Frost, who founded the Medico-Botanical Society in 1821, and who, when his affairs went awry in the early 1830s, fled to Berlin and there practised as a physician. All these men lectured not only to hospitals and associated medical organizations, but to Mechanics' Institutions, the Royal and London Institutions, and/or similar bodies. Burnett, Lindley, Grant and Jones were Professors at University or King's Colleges.

Lectures on biological subjects could draw on another London resource in addition to the talent of the medical community. They could exploit London's position as the centre of entertainment, spectacle and display. London was the home of spectacular animal menageries, such as the Exeter Change exhibit. Private ventures of that type were institutionalized in the 1820s too, with the formation of the Zoological Society (1826) and its gardens.[77] In London were found exhibitions of natural history and stuffed specimens: Bullock's collection, Donovan's collection, and ultimately the natural history department of the British Museum.[78] Phrenology and mesmerism proved fertile fields for exploitation by popular lecturers on the fringes of medicine.[79] In the years after 1825 the popular anatomical waxworks reappeared, blending medical education and thinly veiled eroticism. Joseph Kahn's anatomical museum, the apex of this form of entertainment, exploited still another resource: the oxy-hydrogen microscope.[80] Since that device combined new technology, biological investigation, and visual display, it might be called the perfectly representative London scientific tool. The oxy-hydrogen microscope was first displayed in 1833, and its use proliferated in institutional lecture halls in the 1830s and 1840s.[81] Private lecturers employed it too: in 1833 Rowland Detrosier wished that he had the device, for with it, 'Cooper and Cary have been clearing £10 daily for some time past.'[82]

London biological lecturing, then, could partake of both medical and popular traditions, and it could utilize spectacular advances in technology. It is no less true – though perhaps not so obviously demonstrable – that London scientific lectures reflected its position

in the world of technology. The dramatic advances of industrialization – steam engines, textile machinery, iron-founding, railways,
coal mines – are more often associated with provincial Britain in the
eighteenth and early nineteenth centuries. Such an association may
overlook the vitality of the metropolis in the technological realm.
London had been pre-eminent in the technology of the pre-
industrial revolution period, as the centre of shipbuilding,
instrument- and watch-making, and brewing. In the arts of surveying and instrument-making, of 135 addresses given by Taylor for
the period 1800–9, ninety-one are London; for 1835–40, the figures
are 111 of 192.[83] A quick survey of patentees listed in a few sample
years of *Annals of Philosophy* suggests London's continuing
importance as a centre of inventiveness. The first three volumes of
that journal list seventy patentees with London addresses whose
patents were received in 1812 and 1813.[84] The social variety of the
patentees was wide: six Londoners described as "gentlemen" and
two "esquires", as well as engineers, joiners, merchants, ironmongers, musical instrument makers, a land agent, an architect, a
veterinary surgeon, and an army officer. However the true variety
lay not in class but in technological interests. The list of patents
embodied improvements in sugar-refining, printing, umbrellas,
carriages, musical instruments, chronometers, locks, industrial
chemistry, surveying instruments, military gear, and ship construction and equipment – all areas of traditional importance in the
complex London economy.

In 1823 *Annals of Philosophy* listed a total of 137 patents issued:
of these sixty-six were issued to Londoners, and again the variety of
the patentees and their patents is striking.[85] If any categories are
dominant they are in the general areas of textile machinery and
clothing devices, and improvements in shipbuilding and the making
of naval stores; but printing, locks, musical instruments, military
gear and carriages are still represented, as are inventions relating to
gas-lighting. Some inventions are difficult to classify: for example,
the inspiration of J. Hughes, a Barking slopseller, for a 'means of
securing the bodies of the dead in coffins' – whether against
resurrections from within or without is not clear.[86]

Such evidence suggests that curiosity about technological improvements was extensive in London across a fairly wide social
spectrum. It also suggests that while some traditional industries
remained important in London, that vast city encompassed interests
of extraordinary diversity. London audiences were attuned to

technology and London lectures on the sciences often concentrated on diverse technological applications. Thus the steam engine was discussed in lectures on heat and light; compasses in courses on magnetism; artificial lighting, including safety lamps and lime lights, in courses on heat and light, electricity and magnetism, and 'chemistry as applied to the arts'.[87] Pumps, wells, syphons, and Bramah's press were logical subjects of lectures on the 'properties of matter', on pneumatics, and on hydrostatics and hydro-dynamics.[88] The new subject of photography found homes in lectures on the 'chemistry of heat and light', on the 'correlation of physical forces', on the 'chemistry of the metals', and on the 'chemical agency of the solar rays'.[89] At the London Institution, for example, the appeal of technology was not so much directed to particular London technologies (instrument-making, printing, brewing, shipping) as to exposing the listeners to the wonder of the moment, and in that respect its calendar of Friday evening speakers was remarkably timely: railways in 1832, 1833, and 1836; the oxy-hydrogen microscope in 1834, steam printing in 1837, the electric telegraph in 1839, the daguerreotype in 1840, explosives in 1848, the Conway and Britannia Bridges in 1850, the Crystal Palace in 1851.[90] In each case the lectures quickly followed some striking new development.

Something of the same experience was repeated at the London Mechanics' Institution as well: a large number of lectures on industries and processes, related to scientific principles at times, but not noticeably linked to particular London interests. There were numerous lectures on the steam engine and its applications, especially for transport by land and sea; lectures on navigation, the stability of floating bodies, the preservation of timber from decay, the tides, and the 'means of saving lives by ship wrecks' [*sic*] recall London's position in the maritime world. Other lectures seem directed to London's immense artisan population, working in the clothing and building trades: the manufacture of linens, woollens, lace, paper, quills, steel pens, candles, pottery, and jewellery were all lecture subjects in the early years of the Mechanics' Institution, as were architecture, the strength of materials, chimneys and artificial illumination. But these lectures are simply lost in a sea of others: mining, cholera (in 1832), comets (in 1835), Babbage's calculating machines, the relationship between science and agricul-ture, a series of annual lectures on 'inventions recently patented', and poisons.[91] It would be most difficult to demonstrate that lectures at the Mechanics' Institution adhered particularly to a "London" pattern – beyond repeating that London had a great reservoir of people interested in a wide range of technological and industrial applications.

Lectures on such subjects were available from two different, although somewhat overlapping, categories of lecturers. There was on the one hand that group of men of science who made careers of delivering lectures on subjects which joined science and technology: John Millington, Charles Frederick Partington, Edward William Brayley, William Thomas Brande, William Ritchie, Dionysius Lardner are all good examples. Millington was an engineer, Brande a chemist, Lardner and Ritchie both clergymen-turned-natural philosophy lecturers, Partington and Brayley popular lecturers and writers. Lectures from this group were supplemented by the offerings of more "practical" men who carried on a trade or a business and occasionally lectured on it: John Taylor (a mining engineer), Edward Dent (a clockmaker), Edward Cowper (the inventor of printing machinery), Henry Samuel Boase (the Manager of a bleaching works), Apsley Pellatt (a glass manufacturer), a Mr Stone (master shipwright at the Deptford naval dockyard), a Mr Gutteridge (an instrument-maker). More systematic research into the lecturers at Mechanics' Institutes would undoubtedly produce more such figures, drawn into the London institutional lecturing scene in the 1820s and 1830s. The list of London lecturers compiled in the early 1840s by the Society for the Diffusion of Useful Knowledge contains about 100 names, and it is far from complete.[92]

It was the class of career lecturers – most of them technological, such as Lardner and Ritchie – which made a notable metropolitan impact on the provinces. Few of these men were in the first rank of science as original researchers, yet their provincial appearances, were triumphant. Richard Phillips in Manchester in 1825 excited 'the liveliest interest'; Lardner's appearance in Newcastle in 1836 was 'the grand treat of all', with 'crowded and enthusiastic audiences' which 'handsomely increased' the coffers of the Literary and Philosophical Society.[93] Lardner delivered eighteen lectures in Leicester in 1835 and received £140 12s. for them.[94] Partington was the highest paid lecturer engaged by the Stalybridge Mechanics' Institution in its early years.[95] Robert Addams and Brayley appeared in Leeds, Ritchie in Liverpool.[96] Metropolitan reputations preceded them. The London focus of such journals as

Philosophical Magazine conferred on them celebrity status; the Society for the Diffusion of Useful Knowledge advertised their services; the lecturing opportunities of London allowed them to perfect techniques that would awe the naive provincials. At least one provincial dignitary (John Marshall Jnr, of Leeds) feared that local talents would be inadequate and asked Henry Brougham to recommend London lecturers.[97] The provinces might have sustained an active corps of itinerant lecturers, but evidently the London performer, his career assured by the proliferation of metropolitan institutions, held a special appeal. Yet relatively few of the London medical-biological lecturers extended their activities outside the metropolis. Perhaps London medical practice was for them busy enough and successful enough to make provincial lecture tours unprofitable; perhaps provincial scientific society was not short of medical talents. Thackray's analysis of the Manchester Literary and Philosophical Society suggests the latter.[98]

Another London strength, the possibilities of visual display, was exploited by technological lecturers as well as biological, but it was perhaps more difficult to export to the provinces. Visual demonstration of technology perhaps whetted the public appetite for more information; but mere lectures (even when illustrated by working models) had to endure formidable competition, for London was the nation's (perhaps the world's) centre of popular exhibits. Early in the century private collections and displays of mechanical delights were a part of the London scene; such ventures were gradually supplanted by successful corporate institutions, notable among which were the National Gallery of Practical Science, also called the 'Adelaide Gallery' (1832) and its great rival the Polytechnic Institution (1838).[99] In the 1840s these galleries were among London's great attractions, with their steam guns, model canals, and electric eels only the most sensational of their wide variety of exhibits. The Crystal Palace Exhibition was from this perspective simply the climax of a long tradition of London technological showmanship. This showmanship was difficult for provincials to emulate. It is true that William Sturgeon, who had considerable London lecturing experience, founded the Royal Victoria Gallery in Manchester on the model of the Adelaide Gallery.[100] But generally the railways made it easier to bring provincials to London than to send London technological shows to the provinces; indeed, it has become a cliché that masses of Englishmen and Englishwomen visited their capital for the first time in 1851.

V

London scientific institutions emphasized lectures on technology that utilized visual display and attempted to stay abreast of new developments. London was a city with a diverse and continuing tradition of technological inventiveness and skill. That expertise could be tapped for lecturing talent; that tradition had created a wide audience for the achievements of technology. The institutionalization of the scientific lecture in the 1820s and 1830s contributed to the professionalization of men of science, in that it stabilized their incomes; it also advanced provincial opportunities for metropolitan men. Institutional demands of solvency often drove the presentation of science into smaller and more specific boxes. Biological and technological subjects were ideally fitted to the smaller and more specific box, since they could be subdivided without fear of loss of that continuity which subjects requiring prerequisite knowledge demanded. The small box could be conveniently shipped to the provinces by a lecturer whose reputation was strengthened by metropolitan media and whose career was formed by metropolitan institutions. London's particular strengths – its size, its medical community, its immensely complex technological culture, and its place as the national capital of amusement – made lectures on biological and technological subjects logical for it, and its technology (if not biology) lectures reached out to the provinces.

But in a materialistic age, when as Richard Altick observes 'interest was concentrated upon physical properties and processes' rather than the abstract truths which underlay them,[101] the London scientific lecturer could not rival the exhibition. In the pursuit and exposition of abstract truths the lecturing institution was equally awkwardly placed, especially as professionalization demanded demonstrated competence of the sort which a university degree might offer. London scientific lecturing teetered uneasily between the Crystal Palace and the University of London, and when those extremes matured, the proliferation of popular scientific institutions in the metropolis was at an end. In the provinces the London lecturer also faced the growth of universities which were expressions of an emerging, increasingly self-confident, urban culture; with the foundation of the "civic universities" in the years after 1850, one form of provincial dependence on London declined. London's domination would henceforth be expressed in its control of publishing, its professional societies, and its position in the centre of a radial railway network. The scientific lecturer and his institutions, instruments of London cultural imperialism in the first half of the century, fell between the appeals of the exhibition and the university.

Acknowledgements

I am grateful to the American Philosophical Society for a grant which facilitated the research on which this essay is based. For permission to cite material in their care, I thank the Royal Institution, Birkbeck College, and the Guildhall Library, all in London.

Notes and references

1   See R. E. Schofield, *The Lunar Society of Birmingham: A Social History of Provincial Science and Industry in Eighteenth-Century England*, Oxford, 1963; A. Clow and N. L. Clow, *The Chemical Revolution: A Contribution to Social Technology*, London, 1952; A. E. Musson and E. Robinson, *Science and Technology in the Industrial Revolution*, Toronto, 1969; A. Thackray, 'Natural knowledge in cultural context: the Manchester model', *American Historical Review*, 1974, **79**, 672–709; R. H. Kargon, *Science in Victorian Manchester: Enterprise and Expertise*, Baltimore, Md, 1978; S. Shapin, 'The Pottery Philosophical Society, 1819–1835: an examination of the cultural uses of provincial science', *Science Studies*, 1972, **2**, 311–36.

2   E. G. Forbes, *Greenwich Observatory: Origins and Early History (1675–1835)*, London, 1975, pp. 163–6; A. J. Meadows, *Greenwich Observatory: Recent History (1836–1975)*, London, 1975, pp. 1, 8; R. M. MacLeod, 'The Royal Society and the government grant: notes on the administration of scientific research, 1849–1914', *Historical Journal*, 1971, **14**, 323–58; *idem*, 'Science and the Civil List, 1824–1914', *Technology and Society*, 1970, **6**, 47–55.

3   H. B. Jones, *The Royal Institution: Its Founder and its First Professors*, London, 1871; M. Berman, *Social Change and Scientific Organization: The Royal Institution, 1799–1844*, Ithaca, 1978.

4   L. G. Wilson, *Charles Lyell: The Years to 1841, the Revolution in Geology*, New Haven, Conn., 1972, pp. 136–40; J. G. Crowther, *Statesmen of Science*, London, 1965, pp. 75–101.

5   G. N. Clark, *A History of the Royal College of Physicians of London*, 2 vols., London, 1964–6, **ii**; Z. Cope, *The Royal College of Surgeons*

of England: A History, London, 1959; C. Newman, *The Evolution of Medical Education in the Nineteenth Century*, London, 1957; S. W. F. Holloway, 'Medical education in England, 1830–1858: a sociological analysis', *History*, 1964, **49**, 299–324.

6   T. Kelly, *George Birkbeck: Pioneer of Adult Education*, Liverpool, 1957; A. Wood and F. Oldham, *Thomas Young: Natural Philosopher, 1773–1829*, Cambridge, 1954; P. T. Hinde, 'William Hyde Wollaston: the man and his "equivalents"', *Journal of Chemical Education*, 1966, **43**, 673–6.

7   G. J. Whitrow, 'Some prominent personalities in the early history of the Royal Astronomical Society', *Quarterly Journal of the Royal Astronomical Society*, 1970, **11**, 89–104.

8   F. Sheppard, *London 1808–1870: The Infernal Wen*, Berkeley, 1971, pp. 158–87; A. E. Musson and E. Robinson, op. cit. (1), especially pp. 13–26, 51–9, 119–38; *Life of William Allen, with Selections from his Correspondence*, 3 vols., London, 1846.

9   H. H. Bellot, *University College London, 1826–1926*, London, 1929; F. J. C. Hearnshaw, *The Centenary History of King's College London, 1828–1928*, London, 1929.

10   Newman, op. cit. (5), p. 33. Also A. Chaplin, *Medicine in England During the Reign of George III*, London, 1919.

11   A. Hume and A. I. Evans, *The Learned Societies and Printing Clubs of the United Kingdom*, London, 1853; J. W. Hudson, *The History of Adult Education*, London, 1851, p. 228; Kelly, op. cit. (6), pp. 313–14.

12   A. T. Gage, *History of the Linnean Society of London*, London, 1938; H. B. Woodward, *History of the Geological Society of London*, London, 1908; M. J. S. Rudwick, 'The foundation of the Geological Society of London: its scheme for cooperative research and its struggle for independence', *British Journal for the History of Science*, 1963, **1**, 325–57; J. L. E. Dreyer and H. H. Turner, *History of the Royal Astronomical Society, 1820–1920*, London, 1923; P. C. Mitchell, *Centenary History of the Zoological Society of London*, London, 1929.

13   S. Ross, '*Scientist*: the story of a word', *Annals of Science*, 1962, **18**, 65–85.

14   For the Thames Tunnel, R. D. Altick, *The Shows of London*, Cambridge, Mass., 1978, pp. 373–4; for the balloons, *Philosophical Magazine*, 1802, **13**, 192–201; for Wheatstone, *Personal Reminiscences of Sir Frederick Pollock*, 2 vols., London, 1887, **i**, 269; for Owen, ibid., 269–70, H. N. Pym (ed.), *Memories of Old Friends*,

Being Extracts from the Journals and Letters of Caroline Fox from 1835 to 1871, London, 1882, pp. 153, 206, 248, and R. Owen, *The Life of Richard Owen, by his Grandson*, 2 vols., London, 1894, *passim*; for Herschel's telescope, K. C. Balderston (ed.), *Thraliana: The Diary of Mrs Hester Lynch Thrale (later Mrs Piozzi), 1776–1809*, 2nd edn, 2 vols., Oxford, 1951, **ii**, 6–8, and B. Silliman, *A Journal of Travels in England, Holland, and Scotland in the Years 1805 and 1806*, 3rd edn, 3 vols., New Haven, Conn., 1820, **ii**, 130; for Babbage's machines, S. De Morgan, *Memoir of Augustus De Morgan*, London, 1882, p. 89; for the whale, *Philosophical Magazine*, 1808, **33**, 334; for Bentham, Pym, op. cit. (14), p. 222.

15   For Frost and Lardner, articles in *DNB*; for Millington, article in *DAB*; for Smith, Bellot, op. cit. (9), p. 160; for Knox, W. Roughead (ed.), *Burke and Hare*, Edinburgh, 1921, p. 91.

16   I. Inkster, 'Science and society in the Metropolis: a preliminary examination of the social and institutional context of the Askesian Society of London, 1796–1807', *Annals of Science*, 1977, **34**, 1–32.

17   For example, *Philosophical Magazine*, 1811, **37**, 80. Also E. M. McInnes, *St Thomas' Hospital*, London, 1963, pp. 82–6.

18   These examples are drawn from *Philosophical Magazine*, 1801, **8**, 382 (Pearson); ibid., 1801, **11**, 96 (Garnett); ibid., 1804, **18**, 376 (Great Marlborough Street); *Annals of Philosophy*, 1813, **2**, 235 (Windmill Street).

19   *Philosophical Magazine*, 1802, **13**, 308.

20   Inkster, op. cit. (16), pp. 16–18.

21   *Philosophical Magazine*, 1801, **11**, 96.

22   ibid., 1803, **14**, 364–8.

23   ibid., 1802, **12**, 105–6.

24   *Gentleman's Magazine*, 1828, **98**, 377.

25   L. Davidoff, *The Best Circles: Society, Etiquette, and the Season*, London, 1973.

26   For example, the calendar in *Philosophical Magazine*, 1827, new series, **2**, 400, and in subsequent issues of that journal.

27   Based on an examination of Minutes of the Meetings of the Managers of the Royal Institution, MS volumes in the archives of the Royal Institution, London (hereafter cited as RIMM); and on printed handbills of lectures at the Institution, in the Institution's archives (hereafter cited as RI Handbill).

28   J. N. Hays, 'Science in the City: the London Institution, 1819–1840', *British Journal for the History of Science*, 1974, **7**, 146–62; J. C. Cutler, 'The London Institution, 1805–1933', Ph.D thesis, University

of Leicester, 1976.

29 For example, *Philosophical Magazine*, 1824, **60**, 239 (Surrey); ibid., 1829, new series **5**, 76–7 (Russell); ibid., 1827, new series **1**, 199–207, and *Morning Post*, 30 November 1843, p. 3 (Gresham).

30 Kelly, op. cit. (6), pp. 98–9.

31 M. F. Tylecote, *The Mechanics' Institutes of Lancashire and Yorkshire before 1851*, Manchester, 1957, p. 61; Kelly, op. cit. (6), p. 188. More generally, I. Inkster, 'The social context of an educational movement: a revisionist approach to the English mechanics' institutes, 1820–1850', *Oxford Review of Education*, 1976, **2**, 277–307, especially 284–8.

32 Kelly, op. cit. (6), pp. 313–14; Hudson, op. cit. (11), p. 228.

33 For example, *The Times*, 21 September 1827, p. 1; 2 October 1837, p. 1; 3 October 1837, p. 1; 4 October 1837, p. 1. More generally, Newman, op. cit. (5), pp. 113–17.

34 Mitchell, op. cit. (12); A. E. Gunther, *A Century of Zoology at the British Museum Through the Eyes of Two Keepers, 1815–1914*, New York, 1975; Altick, op. cit. (14), pp. 377–89.

35 C. W. New, *The Life of Henry Brougham to 1830*, Oxford, 1961, pp. 347–58; J. N. Hays, 'Science and Brougham's Society', *Annals of Science*, 1964, **20**, 227–41; M. C. Grobel, 'The Society for the Diffusion of Useful Knowledge, 1826–1846', MA thesis, University of London, 1933.

36 M. Peckham, 'Dr Lardner's Cabinet Cyclopedia,' *Papers of the Bibliographical Society of America*, 1951, **45**, 37–58.

37 J. B. Morrell, 'Individualism and the structure of British science in 1830', *Historical Studies in the Physical Sciences*, 1971, **3**, 183–204; L. P. Williams, 'The Royal Society and the founding of the British Association for the Advancement of Science', *Notes and Records of the Royal Society of London*, 1961, **16**, 221–33; W. F. Cannon, 'History in depth: the early Victorian period', *History of science*, 1964, **3**, 20–38; A. D. Orange, 'The British Association for the Advancement of Science: the provincial background', *Science Studies*, 1971, **1**, 315–29; *idem*, 'The idols of the theatre: the British Association and its early critics,' *Annals of Science*, 1975, **32**, 277–94; *idem*, 'The origins of the British Association for the Advancement of Science', *British Journal for the History of Science*, 1972, **6**, 152–76.

38 Annual report of the Committee of Visitors of the Royal Institution (printed handbills in the archives of the Royal Institution, London), 20 April 1811.

39 ibid., 19 April 1816 and 11 April 1825.

40   E. Ironmonger, 'Forgotten worthies of the Royal Institution: (1) William Thomas Brande (1788–1866)', *Proceedings of the Royal Institution of Great Britain*, 1960–1, **38**, 450–61; C. H. Spiers, 'William Thomas Brande, leather expert', *Annals of Science*, 1969, **25**, 179–201.

41   Minutes of the Quarterly General Meetings of the London Mechanics' Institution, MS volumes in the archives of Birkbeck College, London (hereafter cited as MIQGM), 3 March 1830, **1**, 428–41; ibid., 7 March 1832, **2**, 48–61; 4 June 1834, **2**, 214–27; 7 September 1836, **2**, 353–67; ibid., 6 June 1838, **2**, 463–75.

42   For example, Minutes of the Meetings of the Managers of the London Institution, MS 3076, Guildhall Library, London (hereafter cited as LIMM), 10 May 1827; RIMM, 9 January 1826, **7**, 52; *Philosophical Magazine*, 1834, 3rd series, **4**, 151 (Marylebone); *London Mechanics' Register*, 1824–5, **1**, 394 (Spitalfields); ibid., 1826, **4**, 384 (Southwark); *New London Mechanics' Register*, 1827, **1**, 62, and 1828, **2**, 476–7 (Hackney).

43   *London Mechanics' Register*, 1826, **3**, 402 and 409.

44   RIMM, 7 June 1830, **7**, 352; 6 June 1831, **7**, 408; 1 June 1832, **8**, 21; 1 July 1833, **8**, 103; 7 July 1834, **8**, 203; 1 June 1835, **8**, 354; 6 July 1835, **8**, 368; 8 June 1836, **8**, 431–2; 5 June 1837, **8**, 493.

45   *A Descriptive Catalogue of the Lectures Delivered at the London Institution*, London, 1854, pp. 18–19; MIQGM, 6 March 1833, **2**, 120–32; ibid., 4 June 1834, **2**, 214–27; 3 December 1834, **2**, 245–60; 2 September 1835, **2**, 296–306; 1 March 1837, **2**, 389–407; 6 September 1837, **2**, 420–32.

46   *Philosophical Magazine*, 1834, 3rd series, **4**, 151; Bellot, op. cit. (9), pp. 133–4.

47   RIMM, 3 July 1820, **6**, 297; ibid., 3 June 1822, **6**, 358; LIMM, 9 May 1822; MIQGM, 4 March 1829, **1**, 355–70; *Philosophical Magazine*, 1816, **48**, 235.

48   Bellot, op. cit. (9), p. 135.

49   *DAB*

50   LIMM, 8 November 1832.

51   For example, ibid., 10 February 1825 and 12 May 1831.

52   *London Mechanics' Register*, 1825, **1**, 197–9 (London); ibid., pp. 325–32 (Spitalfields); ibid., 1825, **2**, 362 (City of London L and SI); ibid., 1826, **4**, 366 (Hackney); *Philosophical Magazine*, 1821, **58**, 303 (Surrey); ibid., 1829, new series **5**, 76 (Russell).

53   LIMM, 14 August 1834.

54   RI Handbill, 11 May 1839; *A Descriptive Catalogue of the Lectures,*

op. cit. (45), pp. 16, 18, 20–6; *New London Mechanics' Register*, 1828, **2**, 375–9; MIQGM, 3 December 1828, **1**, 336–51; ibid., 1 June 1831, **1**, 420–33; *Philosophical Magazine*, 1829, new series **5**, 76–7; ibid., 1834, 3rd series, **4**, 151; ibid., 1839, 3rd series, **14**, 158–9.

55 ibid., 1834, 3rd series, **4**, 151; *A Descriptive Catalogue*, op. cit. (45), pp. 18–19; *New London Mechanics' Register*, 1827, **1**, 261–7; for example, MIQGM, 3 December 1834, **2**, 245–60, and 1 June 1835, **2**, 283–93; *London Mechanics' Register*, 1826, **3**, 374–5; ibid., 1826, **4**, 412–13.

56 Berman, op. cit. (3); *idem*, 'The early years of the Royal Institution, 1799–1810: a re-evaluation,' *Science Studies*, 1972, **2**, 205–40; S. Shapin and B. Barnes, 'Science, nature and control: interpreting mechanics' institutes', *Social Studies of Science*, 1977, **7**, 31–74. For other views, E. Royle, 'Mechanics' institutes and the working classes, 1840–1860', *Historical Journal*, 1971, **14**, 305–21; I. Inkster, 'Science and the mechanics' institutes, 1820–1850: the case of Sheffield', *Annals of Science*, 1975, **32**, 451–74; *idem*, op. cit. (31); G. Foote, 'Science and its functions in early nineteenth century England', *Osiris*, 1954, **11**, 438–54.

57 For example, an introductory lecture of 1802, in J. Davy (ed.), *The Collected Works of Sir Humphry Davy, Bart.*, 9 vols., London, 1839–40, **ii**, 311–26.

58 For example, *London Mechanics' Register*, 1825, **2**, 181–6.

59 L. P. Williams, *Michael Faraday: A Biography*, New York, 1965, pp. 322–34; R. J. Seeger, 'Michael Faraday and the art of lecturing', *Physics Today*, 1968, **21** (8), 30–8.

60 Jones, op. cit. (3), pp. 174–7; T. Young, *A Course of Lectures on Natural Philosophy and the Mechanical Arts*, 2 vols., London, 1807.

61 Jones, op. cit. (3), pp. 285–92.

62 Hays, op. cit. (28), p. 156.

63 Kelly, op. cit. (6), pp. 76–108; and a survey of quarterly budgets in MIQGM.

64 H. Ward, *History of the Athenaeum, 1824–1925*, London, 1926, p. 39; Bellot, op. cit. (9), pp. 175–6; Hearnshaw, op. cit. (9), p. 118.

65 MIQGM, 1 June 1836, **2**, 338–49; 5 June 1839, **2**, 522–36; *New London Mechanics' Register*, 1827, **1**, 304–5; ibid., 1828, **2**, 219.

66 Hays, op. cit. (28), pp. 156–8.

67 For examples, *A Descriptive Catalogue*, op. cit. (45), pp. 15–21.

68 H. B. Jones, *The Life and Letters of Faraday*, 2 vols., London, 1870, **i**, 385–6.

69 Often reported in *Philosophical Magazine*; for example, ibid., 1826,

67, 223–4, 300–1, 382–3; ibid., 1826, **68**, 61–2. More complete lists in Ledger Book of Attendance at Royal Institution Lectures, MS volume in the archives of the Royal Institution, London.

70   *A Descriptive Catalogue*, op. cit. (45), pp. 17–26.

71   Kelly, op. cit. (6), pp. 112–13.

72   Minutes of the Meetings of the London Mechanics' Institution, MS volumes in the archives of Birkbeck College, London, 2 February 1824, 12 April 1824, 17 May 1824, and 6 September 1824. See also *London Mechanics' Register*, 1824–5, **1**, *passim.*

73   ibid., 1825–6, **3**, *passim*; ibid., 1826, **4**, *passim; New London Mechanics' Register*, 1827, **1**, *passim.*

74   Based on a survey of MIQGM, 1831–40.

75   Other examples: RIMM, 19 February 1816, **6**, 92; ibid., 11 March 1816, **6**, 98; 8 April 1816, **6**, 107–8; 3 June 1816, **6**, 123; H. A. Ormerod, *The Liverpool Royal Institution: A Record and a Retrospect*, Liverpool, 1953, pp. 28–31; R. S. Watson, *The History of the Literary and Philosophical Society of Newcastle-upon-Tyne (1793–1896)*, London, 1897, pp. 238–47 and 251–3; F. B. Lott, *The Story of the Leicester Mechanics' Institute, 1837–1871*, Leicester, 1935, pp. 6–9.

76   Newman, op. cit. (5), *passim.*

77   Altick, op. cit. (14), pp. 307–19.

78   ibid., p. 369; Gunther, op. cit. (34), pp. 64–115.

79   T. M. Parssinen, 'Popular science and society: the phrenology movement in early Victorian Britain', *Journal of Social History*, 1974, **8**, 1–20; *idem*, 'Mesmeric performers', *Victorian Studies*, 1977, **21**, 87–104.

80   Altick, op. cit. (14), pp. 338–42.

81   ibid., pp. 369–71. For examples, *A Descriptive Catalogue*, op. cit. (45), p. 18; MIQGM, 4 December 1833, **2**, 173–86; ibid., 6 June 1838, **2**, 463–75; Ledger Book of Attendance, op. cit. (69), pp. 78–80, 96–100, 113–17.

82   G. A. Williams, *Rowland Detrosier: A Working-Class Infidel, 1800–34* (University of York Borthwick Papers, **28**), York, 1965, p. 36.

83   E. G. R. Taylor, *The Mathematical Practitioners of Hanoverian England, 1714–1840*, Cambridge, 1966, pp. 354–79 and 466–83.

84   *Annals of Philosophy*, 1813, **1**, 76–7, 156–7, 316–17, 396–7, 475; ibid., **2**, 77, 157–8, 317–18; ibid., 1814, **3**, 76–7, 158, 238, 317. Less conveniently in B. Woodcroft, *Alphabetical Index of Patentees of*

Inventions, London, 1854.

85 *Annals of Philosophy*, 1823, **21**, 318, 398, 472; ibid., **22**, 77–8, 158, 237–8, 318, 398, 472; ibid., **23**, 78, 157, 238.

86 ibid., **22**, 398.

87 Printed handbill of London Institution lectures, in the Guildhall Library, London, by W. T. Brande, 1819 (hereafter cited as LI Handbill); RI Handbill, 26 January 1818 (Millington on compasses); LI Handbills, Brande, 1825; Hemming, 1831; Ritchie, 1837; Solly, 1847 (all on artificial lighting).

88 LI Handbills, Partington, 1828; Lardner, 1832; Addams, 1838.

89 LI Handbills, Brande, 1840; Grove, 1844; Solly, 1846; Hunt, 1846.

90 *A Descriptive Catalogue*, op. cit. (45), pp. 17–26.

91 Based on a survey of MIQGM. The quotation is from MIQGM, 7 December 1831, **2**, 20–37.

92 [T. Coates], *Report of the State of Literary, Scientific, and Mechanics' Institutions in England*, London, 1841, pp. 106–12.

93 Tylecote, op. cit. (31), p. 147; Watson, op. cit. (75), pp. 234–5.

94 Lott, op. cit. (75), pp. 6–7.

95 Tylecote, op. cit. (31), p. 244.

96 E. Kitson Clark, *The History of 100 Years of Life of the Leeds Philosophical and Literary Society*, Leeds, 1924, pp. 160–70; Ormerod, op. cit. (75), pp. 15–31.

97 Tylecote, op. cit. (31), p. 94.

98 Thackray, op. cit. (1), pp. 684–5; I. Inkster, 'Marginal men: aspects of the social role of the medical community in Sheffield, 1790–1850', in J. Woodward and D. Richards (eds.), *Health Care and Popular Medicine in Nineteenth Century England*, New York 1977, pp. 128–63.

99 Altick, op. cit. (14), pp. 350–62 and 377–89.

100 Kargon, op. cit. (1), pp. 37–41.

101 Altick, op. cit. (14), p. 368.

4 The British Mineralogical Society: a case study in science and social improvement

Paul Weindling

Many eighteenth-century scientific societies, like political groups at that time, were notoriously fissiparous personal alliances. They contrast with the formalized specialist societies of the nineteenth century. The British Mineralogical Society, 1799–1806 (henceforth BMS) offers insight into this transition. Although short-lived, it was highly organized and intensely active in the pursuit of economically oriented science. Its intention was to survey the nation's mineral resources: it had a national network of correspondents, although its activities were concentrated in London. Members were primarily apothecaries, chemists and pharmaceutical manufacturers, whose occupations involved mineral analysis. Their mineraological investigations are therefore best explained in relation to the prevailing patronage, dissemination, and utility of science in London. Members' affiliations to other scientific, philanthropic, religious and occupational associations also raise general questions about the proliferation of societies, as well as how in this particular case the improvement of mineralogical theory related to technical improvement and expectations of personal advancement.

There have been conflicting interpretations of the BMS. Historians of science have considered it only as a stepping stone to the foundation of the Geological Society in 1807.[1] Constricted by this narrow interpretation of scientific improvement they have been unsympathetic to the BMS's aim of a "mineral history". Plans for survey and analysis of mineral deposits, as well as the close relation of chemical analysis to members' occupations, are dismissed as of little consequence by those historians of the earth sciences characterizing the achievement of English geology as the investigation of stratigraphical succession as indicated by fossils.[2] Porter's account of the making of geology severs connections between the earth sciences and industrialization by arguing that the BMS was a scientific and technological failure.[3]

In contrast certain economic historians have been sympathetic to the Society, although they have been primarily interested in provincial science in the new industrial areas and the application of science to the increase of industrial productivity. Musson and Robinson have commented on the relation of the BMS to manufacturing interests.[4] But with the far-reaching social changes of industrialization, the extent to which the BMS was symptomatic of new social formations and attitudes also merits consideration.[5] Just as philanthropy changed from individual bequests to organized public associations comparable to joint stock companies, so there was increasing collective activity via scientific societies.[6] A problem posed by the BMS is not whether its minuscule collection of members could make a scientific revolution in the nation's massive mining industry, but how mineralogy was developed in a local context. Thus investigation of the intellectual and occupational interests of members, as well as of their changing social status and views, can be justified as having broader historical significance.

London science

The classic survey of eighteenth-century London by George detected a new scientific spirit of humanitarian improvement towards the end of the century. After the pioneering generation of economic historians like George, Buer, and the Hammonds, who were generally concerned with living conditions, improvement has been more narrowly considered in demographic and economic terms.[7] The metropolis retained supremacy with commercial and political power which scientific and technical innovations could reinforce. But despite George's challenging generalization, there was little investigation of the implications of enlightened improvement in terms of living conditions, social structure or diffusion of ideas in London. There was a corresponding lack of interest in such components of London cultural life as booksellers, printers, engravers, lecturers, surveyors, coffee house clubs, personal patronage by the wealthy, collective patronage by the public, licensing of societies by magistrates, and apprenticeship as a means of education.[8] Now, however, many public institutions, like prisons, madhouses, hospitals, dispensaries, schools and scientific bodies like the Askesian Society (f. 1796) and the Royal Institution (f. 1799) (associated with the Board of Agriculture and the Society for the Bettering the Conditions of the Poor) have recently been investigated to ascertain whether they represented either disinterested scientific advance and the amelioration of public welfare, or social control and the self-interest of social upstarts or the propertied establishment. The validity of such distinctions may be evaluated through the BMS, as certain members had close links with the Royal Institution and the Askesian Society.

Undaunted by the sheer lack of knowledge of London's cultural and social structure, the studies of Berman on the Royal Institution and Inkster on the Askesian Society have offered different interpretations of the social function of science in London. Neither of these applies fully to the BMS, although both authors have made a valuable factual and theoretical contribution. Berman presents the Royal Institution as providing an ideology of social improvement for its landowning aristocratic proprietors as an antidote to the French Revolution, and of economic improvement to attract the support of London's mercantile elite.[9] Inkster suggests that science was a means for marginal groups like dissenters and manufacturers to secure social legitimation. The marginality of provincial scientific groups implied the necessity of a 'social image, as the centre of established society, wealth and culture' for which he looked towards London. Finding that 'like the urban provinces London had its marginal groups *within* the city', the Askesian Society was analysed as a prime example of such a marginal group.[10] Whereas Berman describes the Royal Institution's science as an entrepreneurial ideology to defend the social status of a small group of improving landowners, for Inkster scientific culture was itself a dynamic enabling a rise in status for the "marginal men" of the Askesian Society. Whether the latter interpretation also applies to the BMS will be discussed here: it will be suggested that the BMS was not merely an offshoot of the Askesians but had substantially differing membership and aims.

The BMS's organization required considerable scientific and technical expertise. Consequently its members were scientifically superior to the lecture-going amateurs of the Royal Institution. The specialized nature of the BMS was dependent on occupational and technical factors, which in turn were related to social status. Whereas Berman and Inkster lay stress on the power of science to attract patronage over and above its appeal as a technical agent, BMS members possessed the requisite technical skills. They had to be 'able and willing to undertake a chemical analysis of a mineral substance'.[11] Their skills opened new occupational opportunities and made them intermediaries between the leisured scientific elite and practical men like surveyors. This role permits reappraisal of definitions of social status which exclude the prestige attached to wealth and the advantages accruing from superior capabilities.

Scientific activity in London in the 1780s and 1790s was leavened by technological concerns, especially with provincial manufacturing, and by awareness of new theoretical developments, notably from France. An example of such technical interest was the Chapter Coffee House Society, which showed its practical orientation by electing provincial manufacturers as honorary members (Keir, Watt and Boulton). Industrial processes were frequent topics of conversation. The mineralogist Kirwan discussed Swedish iron and steel manufacturing. William Babington, who later associated with the BMS, spoke of visits to factories and quarries – and of the jealousy of owners.[12] Babington combined technical concerns with chemical classification of minerals, as did Bryan Higgins's Society for Philosophical Experiments and Conversations, attended by William Allen a founder of the BMS. Higgins advocated Lavoisier's new system of chemistry and his Society owned a 'systematic cabinet of mineralogy, metallurgy and vitrifications'.[13] In the similarly technically minded London Philosophical Society (f. 1794) two more BMS founders, Tilloch and Pepys, were active. Pepys's early experiments on fermentation and gases made him acquainted with the researches of such intellectual leading lights as Priestley, Lavoisier and Black, as well as bringing him into personal contact with the immensely wealthy chemist Cavendish and the prestigious circle of Sir Joseph Banks, the President of the Royal Society.[14]

This suggests that certain members of the BMS had prosperous backgrounds and were well connected to influential intellectual groups. Their contact with provincials could confirm the thesis of marginality as could their subsequent absorption into the Royal Society. But continual hob-nobbing in relatively exclusive circles, before the foundation of the BMS and throughout its existence, lessens the impression of cultural marginality. 'Marginal to what?' One may ask. If the answer is the Royal Society and Banks (who maintained contacts with such marginal men as the Lunar Society provincial network, land-surveyors and reforming groups of physicians), marginality as a concept, rather than the social conditions it is meant to describe, seems ill defined. The evidence of the BMS suggests that greater importance should be attached to the persistence of occupational interests and family influence as a basis for contact between scientific groups. It is not enough to take the formation of scientific interests as given and to use only intellectual associates as an index of social mobility.

Conditions in London were favourable for the BMS as there were growing opportunities for earning a livelihood from science. Members' occupations suggest the technical concerns of the Society (see Table 1, p. 138). An important bond between members was pharmacy; indeed the Society met at the Plough Court pharmacy of William Allen. The members also engaged in lecturing and publishing, which provided a profitable means for the diffusion of their researches. Arthur Aikin became a prolific editor and lectured with his brother on chemistry and chemical manufactures. Tilloch was an inventor of new printing methods, editor of the evening paper *The Star*, and in 1797 founded the *Philosophical Magazine*, which regularly carried communications from the BMS. Babington considered in his chemical lectures at Guy's Hospital that occupations and science were closely connected: should a potter, tinker or dyer 'reflect and examine the principles on which their respective callings are founded, they then become men of science', each having a duty 'to make general applications of it to the benefit of his fellow creatures'.[15] Babington had been responsible for manufacturing all the Hospital's drugs free of impurities.[16] Allen and Luke Howard (of the Askesian Society) undertook not only scientific experiments but also large-scale manufacture of chemicals.[17] The expectation of financial gain could be a major consideration in the choice of scientific activity: thus Arthur Aikin abandoned research on optical glass owing to punitive excise regulations.[18] Perceptions of the connection between science and technical application differed, but there were clearly considerable expectations about the benefits of applied mineralogy. As members shared common scientific interests, it may of course be objected that it is irrelevant to look beyond a short-lived fashion for chemical mineralogy as the basis of the Society. But occupations provided the necessary scientific training and facilities.

Technical priorities were to the fore in the Society's interest in French chemical theory and the Paris School of Mines. The *Monthly Magazine* (edited by John Aikin, the father of two BMS members) commented on French technical schools:

This interest is doubly enhanced by the consideration, that the real state of the country, and its improvements, from some sage motives of bottle conjuring policy, are attempted to be sedulously withheld from the observation of the people of this country.[19]

It was therefore hoped that the lack of 'any college or school of scientific mineralogy' in Britain was to be remedied by members' collective abilities as 'philosophical chemists'. That the French Revolution had been attributed to philosophical materialism put men of science in a vulnerable position. But economic utility was more than a convenient public front. Although the mining industry was in economic difficulties, the challenge of war and the need to provide employment showed the desirability of applying science to mining and metallurgy.

The organization of the BMS

The intention of co-operative research was successfully realized. Meetings were held every fortnight; a fine was imposed for absence when it was a member's turn to analyse a specimen. Between January 1800, when the organization was finalized, and January 1804, when the scientific aims and meeting place changed, seven members on average attended each session. The highest participation was achieved by Pepys, who was commended 'for the active and disinterested attention to the Interests of the Society which he has uniformly shown from the moment of its foundation'.[20] Those who exceeded Tilloch's attendance of 37 per cent of all meetings were Allen, Knight, A. Aikin and C. R. Aikin, Laws, T. Cox, Oldham, R. Phillips, Sandman and Bingley (see Table 2, p. 141). This active nucleus had no unifying religious or political characteristics, but they shared scientific and occupational interests in chemical mineralogy.

It is important to correct the impression that the BMS was only a specialized group of the Askesian Society. This has led to misconceptions about their origins and composition. Rather than characterizing the Askesian as a large informal association, a summary of the Minutes suggests it was more formal and had a smaller core of active members than Inkster has allowed, although it certainly attracted many visitors (see Table 3, p. 143). Only Allen was a founder of the two Societies; only Tilloch, Pepys, Babington and Tupper were members of both before the Societies amalgamated in 1806. Most of the shared members joined the Mineralogical Society first. The Aikins (although A. Aikin was a visitor), Lowry and

Knight were never Askesians. The two Societies differed considerably in ethos: all but one of the Askesian founders were Quakers, but no religious affiliation predominated in the BMS. By interpreting the BMS as only an offshoot of the Askesians, Porter reduces the importance of occupational factors in favour of a theory of scientific specialization that general societies like the Royal Society or the Askesian inevitably precede specialized societies like the Geological or BMS.[21]

Since it was intended to undertake a national mineral survey, the BMS had corresponding members in the provinces. In 1800 Richard Kirwan of Dublin, George Overton of the Dowlais Iron Works, and the Unitarians William Henry of Manchester and William Turner of Newcastle were elected. In 1801 C. R. Aikin proposed the Unitarian John Taylor of Tavistock, a mining engineer and entrepreneur. Taylor came from a prosperous Norwich family, and his science and business acumen made him regarded as 'one of the wise men from the East'.[22] Tilloch enlisted David Mushet, who became a noted contributor on iron and steel production to the *Philosophical Magazine*, and who in 1801 discovered black band iron ore. After Mushet's move from the Clyde Iron Works to the Calder Iron Works, John Wilson from the latter also became a corresponding member.[23] The Phillips family had Cornish mining interests and were related to iron manufacturers.[24] Richard Phillips presented the Society with Cornish minerals and ores. In 1800 he proposed the mining entrepreneur John Williams Jnr, who was one of the West Country's leading chemists and crystallographers. William Phillips donated a mineralogical map of Cornwall and silver from Herland mine, in which he had a share. Allen's Quaker companion on botanical fieldtrips, Lewis Dillwyn, was another corresponding member (from 1803 he was to manage a large pottery factory near Swansea). Dr Grant Yeats of Bedford communicated on the mineralogy of his district. John Moody of Birmingham sent specimens illustrative of coal strata; and White Watson, the Derbyshire mineral surveyor, sent limestone for analysis.[25]

Despite many contacts with mining engineers, the BMS failed to recruit William Smith, the mineral surveyor, who was engaged on his pioneering stratigraphical survey of England. Smith was informed that Lowry:

had been out this and the last 2 Summers for several weeks on Mineralogical Tours with Mr Aikin, a skillful Mineralogist and Chemist and that it was their object to try to collect materials for a Mineralogical Map of great Britain.[26]

Although Lowry wished to unite the schemes, Smith's highly developed proprietary interests prevented this.[27] In any event Smith's approach to fossils was markedly different from the BMS's chemical methods. Although there were rival plans for national surveys, no acrimonious public debate resulted as when Smith accused the Geological Society of plagiarism. The BMS's correspondents did however anticipate the network-research of the Geological Society. It was suggested by the surveyor, B. Bevan, that the BMS participate in a Society for Scientific Information with a national network of correspondents.[28] That the majority of the BMS's correspondents were surveyors, engineers or manufacturers indicates its occupational orientation. But this meant that contacts were limited to mining areas.

Scientific activities

BMS members were not just closeted in their cabinets and laboratories. They undertook substantial tours and fieldwork to ascertain not only existing mineral resources, but also the processes by which mineral deposits had been formed. The Aikins followed in the footsteps of their father who had toured to acquire historical information. Arthur Aikin toured Wales and Shropshire, where he had been a minister, with the aims of observing strata and, 'To see the whole process also of mining; of extracting the ore, reducing, refining, and manufacturing it'.[29] There was a summer recess so members could tour. Allen, already a devotee of botanical fieldtrips, went to Cornwall in 1804 and observed mines as at Dolcoath. In 1806 he toured Wales and in 1807 visited Cumberland and Yorkshire with Luke Howard, paying special regard to mines and manufactures.[30] A dictionary of local mining terms was planned. Not only did the Society acquire a mineralogical collection, but members had private collections. The Swedish mining engineer, Swedenstierna, commended the Cornish copper specimens in R. Phillips's cabinet and the remarkable Irish minerals in Babington's collection.[31] There was a considerable appetite for new finds of minerals. Greenough complained during his tour of Scotland in 1805:

The supply of Strontian is by no means equal to the demand. The finest pieces yet found were dug up last year and were purchased immed. by Davy and Allen who have drawn upon themselves some censure by the exorbitant prices they paid for them.[32]

In response to the Society's offer to analyse specimens sent in by the public, by 1804 twenty-one specimens were received and the results published. The analyses of calcspar and satinspar were of enduring scientific value.[33] One must qualify the view that the BMS failed because of its '*passive* task of analysing specimens sent up casually from the country. Few specimens were sent in and interest tailed off'.[34]

As an occupation mineral analysis meant economically profitable routine operations; as a science it was the pathway to new discoveries when the laws of chemical combination were unknown. Much attention was paid especially by Pepys, A. Aikin and Knight to the design of apparatus like furnaces, blowpipes, gas-holders, and batteries.[35] The Society's interests ranged over the nature and effects of acids, galvanism and oxidation on minerals and their use in manufactures. Reagents were prepared by a committee to prevent errors arising from impurities. Members collaborated in experiments. In October 1801 it was decided to test the effect of "oxygen gas" on earths and metals: Pepys undertook this for iron, Allen for copper, Tilloch for antimony, Phillips for lead, and Sandman for silver. Experiments on minerals in states of decomposition were intended to show how they had been formed.

Complementing collective experimentation was the Society's keen attention to the monthly journals. Transmission, discussion and experimental investigation of news were seen as essential aspects of mineral history. Economic applications of chemistry in mining, metallurgy and agriculture were discussed, as was the manufacturing of porcelain, candles, pottery, soap, vegetable dyes, and paper. Pepys considered the welding of iron and steel, and showed specimens of "Indian steel" or "wootz", to which Mushet and Banks had been devoting attention. Humphry Davy communicated to the BMS on tanning, a concern of the Royal Institution due to the economic motives of manufacturers and as a means of employing the rural poor.

After the *Monthly Magazine* of May 1801 announced that the BMS was prepared to analyse soils, agricultural matters were discussed in meetings.[36] A. Aikin and Tilloch considered the germination of seeds, frosted potatoes, agricultural mineralogy and an agricultural college. At this time Davy was investigating agricultural chemistry for the Royal Institution and Board of Agriculture. Agricultural mineralogy fitted in well with the BMS's plans for more efficient use of national resources, and offered occupational opportunities: Lynam performed analyses for the Board of Agriculture; Davy advertised equipment supplied by Knight and recommended Allen to the agriculturalist, Arthur Young, as a skilled analyst.[37] The geological relevance of the related field of animal chemistry was shown by analysis of bone: Babington commented that Pacific islands were made of lime of animal origin, the greater part of the earth being composed of calcarous matter.[38]

Investigation of the nature, production and use of ores and new metals like titanium, tantalite, wolfram and columbium also shows the combination of technical and scientific concerns in the BMS. A striking example was members' research on platinum, a metal both profitable and scientifically controversial. Allen fused platinum with oxygen and charcoal in 1799. Knight and Tilloch in 1800, and Bingley in 1802 all devised methods of rendering platinum malleable. But it was due to Thomas Cock, a "gentleman chemist" that a process for the commercial production of platinum objects was discovered; and Cock's brother-in-law, Percival Johnson, later put this into operation. Allen produced articles in platinum by 1805 when he conducted experiments at Plough Court with Cock, one of the Howards, Davy, R. Phillips and W. Henry. The anonymous public announcement (by Wollaston) in 1802 of the discovery of a new element "palladium" provided added stimulus to experiments on platinum, as this was considered by some as a hoax. Pepys tried to form palladium with a powerful battery, and also began to manufacture platinum cutlery by 1805. In addition to the wealth he inherited, Wollaston made a fortune from manufacturing platinum articles. The technical achievement of members was indicated by Boulton in 1806: he considered 'the refining and perfecting of that most obstinate metal Platinum of vast importance to the arts'.[39]

Contacts and competition with the Royal Institution

The Royal Institution came to provide an alternative outlet for the energies of certain BMS members. Davy was not only a key figure at the Royal Institution, but he formed close friendships with Allen, Babington and Pepys, visited the BMS and influenced its scientific activities. Davy's opposition to caloric (the concept of heat as a material substance) and to Lavoisier's oxygen theory of acidity as the basis of all matter can be seen as raising fundamental problems investigated by BMS members' experiments on respiration, galvanism, combustion and oxides. Davy's interest in formative forces which had shaped the earth, like electricity, was associated with new theories of the composition of matter which have been characterized as vitalist.[40] Whether such theoretical changes would have made chemistry less suspect as a vehicle of atheist materialism is a moot point, especially with regard to the controversy about the extent to which Davy was the ideologist of an aristocratic elite, anxious to uphold the existing social order.[41]

Increasing mineralogical activity at the Royal Institution resulted in its superseding the BMS's scheme for public mineral analysis. Pepys was a member of the Institution's committee for chemistry and investigated alloys with Davy.[42] Their activities were discussed in the BMS on 21 January 1802. Davy built up a geographically arranged mineral collection for the Institution after tours of Cornwall and Ireland. The Institution held out the promise from 1804 to 1806 of the foundation of a National Collection of Mineralogy and School of Mines. Davy offered to undertake analysis of minerals and soils as a public service.[43] Simultaneously the BMS deliberately abdicated its responsibilities: in January 1804 it was resolved to discontinue regular meetings.

Behind the scenes a new private group formed as regular *conversazioni* began at Babington's to which the apparatus and minerals of the BMS and Askesians had been removed.[44] An incentive was instruction in crystallography given by Count Bournon, a French émigré who also figured prominently in the Royal Institution's scheme for a national mineralogical survey. The group was joined by Davy, Greenough, William Phillips, Dr Franck and Dr Laird. After the Royal Institution scheme had failed to materialize, they founded the Geological Society in 1807.[45]

If by 1804 new interests and acquaintances meant the BMS had outlived its usefulness, it is still simplistic to write the Society off as having failed. Its scientific achievement had been considerable, its organization provided a model for subsequent schemes and mineralogy acquired public prestige. This growth – rather than decline – of interest in mineralogy meant more extensive laboratory facilities were required than could be provided by the BMS. An attempt was

made to continue the BMS in the City with the founding of the
London Institution in 1805:

The advantages to the Institution with the formation of a School of
Mineralogy and Collection of working and scientific specimens would be
considerably enhanced by accommodating the Mineralogical Society.

It was hoped:

the repet[it]ion of Askesian Experiments in the laboratory with the
opportunity authors might have of elucidating their papers would be
mutual – the lectures would by these means have a fair chance of
introducing much new matter.[46]

Most subscribers to the London Institution were City tradesmen
with comparable occupations to the Askesian and Mineralogical
Societies' members. Samuel Woods was Secretary of the Institution
and Pepys designed the laboratory on which several thousand
pounds were expended.[47] The Askesian minutes noted:

A London Institution having been formed and most of the Members of the
Two Societys having become Members, It was deemed not requisite to
continue the Societys meetings, till further notice be given.[48]

But public patronage was a risky undertaking, for when the
Institution ran into financial difficulties, it was decided to curtail
expensive original research and restrict activities to popular lectures.

Comparable limitations on research occurred with the failure of
the Royal Institution proprietors to support a School of Mines, and
later in the Geological Society where the hope of a national survey
for industrial ends was also buried. This coincided with a shift away
from paternalist economic and philanthropic schemes to a new
emphasis on the benefits of individual private enterprise. Interest in
mineral history waned. In 1816 William Phillips warned the
Geological Society that mineralogy was not just the collecting of
rare specimens but an essential prerequisite for scientific geology,
and Arthur Aikin criticized the ignorance of many geologists.[49]

In both the Royal and London Institutions there was an endemic
conflict of interests: proprietors were anxious to curb the profligacy
of the active scientific nucleus. Berman's analysis of the Royal
Institution in terms of landed aristocracy and City merchants has
not taken levels of scientific expertise adequately into account and

has consequently overlooked the importance of the scientific nucleus as an interest group.[50] The scientific committees were socially more heterogeneous than the proprietors, with City manufacturers like Allen and Pepys as active participants. An outlet for the tension between expert and patron was the growth of further informal scientific groups, supported by experts' independent means.

Scientific societies and social status

The concept of marginality, describing the social circumstances of scientists, has been used to account for an apparent upsurge of scientific activity as a channel of social mobility. On the assumption that there was no clear social apex during the period 1790–1810, Inkster explains the marginality of the Askesians in terms of cultural and institutional mobility; for him socio-economic factors like wealth and occupational bodies like Livery Companies were less significant indicators of social status. While hitherto general historians have neglected science as an important form of association in this period, it is an exaggeration to claim the Askesians were in the metropolitan context "new men" who had abandoned their families' established interests.[51] The concept of marginality tends to over-generalize the importance of science as a means of social mobility. It is possible to interpret scientific activities as maintaining members' family and broader social interests in response to political unrest and economic change.

For the Askesian and Mineralogical Societies, the reduction of different sectarian and political affiliations to the common denominator of marginality does not take account of the persistence of religious, political and professional distinctions which collective scientific activity did not overcome. Inkster considers both Societies equally marginal. I would suggest they differed: the Askesian was a general scientific society dominated by wealthy Quakers, and the Mineralogical Society was preoccupied with specialized science related to occupational concerns.

I would agree with Inkster's intention of evaluating the status of men of science 'as defined *within* the city', but not with his criteria which take no account of wealth and consequent social status in a commercial environment where tradesmen were a substantial political force. Historians' investigations of the distribution of wealth in London suggest that BMS members were of relatively high social standing. Rogers observes increased self-recruitment within London's civic elite, resulting in permanent dynasties. Occupation in itself may be a misleading guide to social status. From 1765 to 1774, for example, 40 per cent of apothecaries' apprentices were sons of esquires or gentlemen. Rubinstein has demonstrated that London as a whole remained more productive of wealth than the new industrial areas, and that the metropolis had a distinctive social structure with Anglican-dominated commercial and professional communities.[52]

By overlooking wealth, Inkster has tended to reduce the status and heighten the marginality of the Askesians. Inkster considers that the Askesians broke with their family traditions and lived on slender incomes.[53] This interpretation is rather over-individualistic as ties of patronage, kin, professional corporations and religion continued to be rife and there was an increasing sense of common interest among the propertied. In order to demonstrate that many differentiating social factors are obscured by the assumption that dissenters, radicals, provincials or such a varied category as medical practitioners were all marginal, crucial characteristics of the BMS and Askesian membership will be considered: family background, recruitment into the Royal Society, and political attitudes.

The Askesian Society, dominated by Quakers, had a smaller, and socially more exclusive core of membership. Although Quakers were subject to civil disabilities, within the City this group was hardly marginal as the wealth and status of the families were well established. Quakers could be accepted members of Livery Companies: for example the Howards and Bevans of Plough Court were influential members of the Ironmongers Company.[54] Luke Howard's father was a prosperous tin-plate manufacturer of Clerkenwell who settled £10,000 on his sons. Luke married a wealthy heiress of the Eliot family of City merchants in 1800. If by 1806 Luke's chemical factory had a turnover of £6000 and by 1825 of £55,000, this was the successful maintenance of his family's status.[55] Henry Lawson and the Woods brothers had considerable independent means. The family of Joseph Fox (of Guy's) included generations of medical practitioners; he was able to subsidize the educational schemes of Joseph Lancaster with several thousand pounds.[56] Besides mining interests, the Phillips family had country property. While Davy commented that Pepys alienated many with his citizen's manners, his status within the City was also secure. His grandfather had been Master of the Clothworkers

Company, and like his father, who was Master of the Cutlers Company in 1793, he was Master in 1821.[57] Lowry's children became engravers and his father had been a portrait painter.

These lineages may be compared to the social origins of medical practitioners in nineteenth-century London. Peterson has shown the highest proportion were themselves sons of medical men.[58] Nepotism flourished at Guy's, with the dynasties of Cooper and Babington.[59] Although both Astley Cooper and Babington came from the provinces, Cooper's appointment was due to family connections and Babington probably benefited from rents of Irish estates so that, even when an apothecary, he acquired the mineral collection of the Marquis of Bute. Babington's son became physician at Guy's and his daughter married another Guy's man, Richard Bright, the son of a wealthy Bristol banker. Astley Cooper profitably married the daughter of Thomas Cock's father, who was a wealthy merchant. The Stockers, too, were a noted family of apothecaries, and Richard Stocker was succeeded at Guy's by his son. One may question the extent to which such figures were "new men" in the 1800s. In cases of differing parental occupation, broader social commitments could still be maintained. Although Allen's father was a silk manufacturer, his sons continued Quaker traditions.

Wealth was also a prerequisite for the ambitious scale of the BMS's activities. Experiments used costly materials like platinum, silver and gold wire or diamonds.[60] As members' occupations provided the necessary private laboratory facilities, it is not always possible to demarcate between the activities of individuals and the Society. Mineralogy was both a luxury and a practical investment which reinforced advantages due to family connections and occupational interests. Bingley had a profitable private assaying practice and he was succeeded by his son as Assay Master at the Mint.[61] Babington, Tupper and C. R. Aikin were to make successful careers in medicine and surgery. Moreover Babington and Aikin were leading members of the Medical and Chirurgical Society (f. 1805) which challenged the authority of the Royal College of Physicians. Pepys adopted a similarly reformist stance: he led the agitation of the Cutlers Company for free trade against the Sheffield monopoly.[62] Allen and R. Phillips used their analytical skills in association with Stocker to criticize the *London Pharmacopoeia* and the monopoly of the College of Physicians. The Apothecaries Act of 1815, for which Allen agitated, brought considerable benefits, as it stipulated compulsory attendance at chemical lectures. Phillips then ceased to practise as a pharmacist and became lecturer in chemistry at the London Hospital.[63] The Act was meant to raise the competence of the medical profession, but as Berman observes, it reinforced the perception that improved status could only be justified in terms of the mastery of scientific theory.[64] The chemical lectures given at Guy's by Allen, Babington, and later by A. Aikin, contained material not only relating to medical theory and dispensing, but were also courses of general and applied science with considerable reference to industrial processes.[65] This suggests that the BMS saw science as having practical occupational value and as a reformist movement to enhance professional success. To dismiss the BMS as underlabourers of a wealthy, leisured 'dominant, innovative, geological elite' fails to recognize adequately its scientific contribution and professional aims.[66]

Although certain common social and scientific interests gave scientific groups homogeneity, in other respects the scientific community was by no means united. Several BMS members became Fellows of the Royal Society, but of the most active in the BMS, Tilloch, the Aikins, T. Cox, Oldham, Sandman and Knight did not. Allen only became FRS after the election of his associates, A. Cooper (1802), Davy (1803), Dillwyn (1804) and Babington (1805), could support his election.[67] Pepys was then elected in 1808, Bingley in 1809, Lowry in 1812, R. Phillips in 1822, W. Phillips in 1827 and Tupper in 1835. The election of Tilloch was opposed as he was editor of a newspaper and thus drawn into "the political vortex". The Aikins were also active in journalism and had until at least 1802 radical sympathies.[68] Banks, the President of the Royal Society, had a hierarchical view of the scientific order: specialist societies were to be dependent on the Royal Society, and the rule was "no politics". He hoped science would yield economic benefits to strengthen the existing social order. He resented the London Chemical Society of 1806 as undermining the authority of the Royal Society. When the Geological Society rebelled against this, he looked to Davy, Babington and Allen for support.[69]

Philanthropists not only tended to see social problems as technical issues requiring applied science, but were also inspired by scientific concepts. Philanthropy provides an acid test for the social outlook of certain leading Askesian and BMS members. Inkster has suggested that the Askesian Society had a 'political profile' with 'an informal network between the Society and radical and political leaders'.[70] Allen, Fox (of Guy's) and W. Phillips were especially active as organizers of charitable schemes in Spitalfields, and other members subscribed. But while the Quaker humanitarian ethic aroused the philanthropic potential of science, it often curbed explicit political activity. Allen admired Wilberforce and Pitt, leading opponents of radicalism. McCann has convincingly argued that charitable and educational enterprise in Spitalfields was repressive and counter-revolutionary; Allen used his connections with families engaged in brewing and banking to obtain support.[71] The Spitalfields Soup Society, which dispensed soup made according to the scientific calculations of Rumford, required the personal attendance of the committee. Participation in such associations formed ties between privileged groups among whom there had been little contact prior to the revolutionary threats of food riots. Quakers were accused by rioters of causing the high price of corn. Quakers' prominence in manufacturing and commerce was defended by philanthropy and extended by science. Scientific theory also provided the attraction of devising cheap and efficient forms of charity like soup kitchens: botanical and chemical experiments even suggested water was nutritious.[72] Benthamites had to counter Fox's insistence that the Bible was to be the basis of all educational instruction, and when a partner in New Lanark Allen was to insist that Owen retain the use of the Scriptures. Such confrontations suggest that science could be a means of countering radicalism.[73]

Askesians were not involved in radical politics. Cooper had abandoned his radical associates before his brief membership. Apart from the radical sympathies of the Aikins, Lowry and possibly Tilloch, only one BMS member was an active radical. Robert Albion Cox was a radical leader of the City livery men, and was Recorder at the Westminster election of 1802 when Burdett was elected. Cox was accused of corruptly accepting votes based on spurious qualifications and was briefly imprisoned.[74] The heterogeneity of the BMS members in their politics confirms that the primary unifying factor for their mineralogical endeavours was originally occupational.

Conclusion

The linear assumption of scientific development from general to specific scientific societies is simplistic. In the case of the BMS a reverse process occurred with occupational interests at the root of the growth of interest in mineralogy. Members then participated in more general scientific associations. Not only were the scientific ventures of the BMS considerable, but members were to be active in more general scientific associations like the Royal and London Institutions and the Geological Society. The chronology of how scientific interests and personal associations developed reveals that important features of geology like touring were established features of mineral history.

The branching-tree model of the progressive differentiation of sciences obscures intellectual interests and prevents examination of socio-economic factors. Just as it is mistaken to evaluate mineral history purely in terms of relevance to subsequent geological theory, so members' social status should not be judged solely in terms of subsequent career. This criterion heightens the impression of social mobility from marginal status. It cannot be assumed that there was inherent marginality to dissent, politics or to youths on the threshold of their careers. Families of many Askesian and BMS members were well established in the City. Scientific training was provided in part by apprenticeship for which family connections were important. Mineralogy provided a means of adapting traditional occupations like those of druggist, cutler, ironmonger or printer to changing socio-economic circumstances rather than merely legitimating marginal, socially mobile groups. This is not to deny that science could confer respectability or provide a means of social mobility; but the formal organization of science could also restrict such mobility. Science could be used by well-established groups to maintain their social standing.

The persistence of family and occupational interests qualifies the generalization implicit in the concept of marginality: that a well-defined social structure broke from the 1790s to the 1820s when social divisions again became apparent. An alternative interpretation is that during these years social distinctions became articulated in class tensions as a response to unprecedented socio-economic change. But just as social mobility in itself does not signify fundamental change of social structure, so broader intellectual and social issues cannot be reduced to terms of just mobility. The making of the middle class in Britain needs investigation; how this occurred economically, politically and culturally in London has hardly been studied. Our knowledge of the context of scientific associations is so inadequate that attempts at generalization must be regarded as extremely tentative. But the case of the BMS suggests

Table 1: *Profile of the British Mineralogical Society, 1799–c. 1806*

Name		Occupation	Religion	Politics	Address
Allen, William	(1770–1843)	Druggist and chemical manufacturer	Quaker	Supporter of Pitt	Plough Court, Lombard Street Poultry
Pepys, William Hasledine	(1775–1856)	Cutler, surgical and scientific instrument maker			
Tilloch, Alexander	(1759–1825)	Publisher and printer	Became Sandemanian (Goswell Street Christians)	Radical?	
Knight, Richard	(d. 1844)	Ironmonger and scientific instrument maker			Foster Lane
Lowry, Wilson	(1762–1824)	Engraver		Lapsed radical	Great Titchfield Street?
Thezard, Alexander Lynam, Charles		Apothecary of Spectacle Makers Co.			Aldermanbury
Aikin, Arthur	(1773–1854)	Editor and chemical lecturer	Temporarily lapsed Unitarian	Lapsed radical	Aldersgate Street, near Foster Lane
Aikin, Charles Rochement	(1775–1847)	Surgeon and chemical lecturer	Unitarian		Aldersgate Street, near Foster Lane
Laws, Charles		Clerk (?) at Bank of England			

Name	Dates	Occupation	Religion		Location
Cox, Theodor Samuel	(d. c. 1803–7)	Of Goldsmiths Company	Anglican		Little Britain, near Smithfield
Phillips, Richard	(1778–1851)	Allen's trainee and later chemist and lecturer	Quaker		George Yard, Lombard Street
Cox, Robert Albion	(d. 1826)	Of Goldsmiths Company and silver refiner	Anglican		Little Britain
Oldham, William Chapman		Apothecary		Radical	Finsbury
Sandman (or Sandemann), Phillip		Chemist and oil of vitriol manufacturer			Southwark
Campbell, Hector		Physician and bleach patentee			Fleet Street and Bermondsey
Coxwell, Henry		Chemist and druggist			Fleet Street
Tupper, Martin	(1780–1844)	Medical student and later physician			Guy's Hospital and matriculated at Oxford 1802
Bingley, Richard		Assay Master at Mint, 1798–1836			Tower of London
Babington, William	(1756–1833)	MD and surgeon (previously apothecary)	Anglican		Guy's Hospital and Basinghall Street, and then Aldermanbury
Stocker, Richard	(1760–1834)	Apothecary			Guy's Hospital
Cock, Thomas	(1787–1842)	Allen's trainee			

Note to Table 1 overleaf

Notes to Table 1

Notes: George Fordyce and John Stancliffe were also proposed as members. For Campbell, see Clow, op. cit. (18), p. 266, and Musson and Robinson, op. cit. (4), pp. 134–5, 334. For Oldham and Coxwell see W. Holden, *Triennial Directory*, 2nd edn, London, 1804. For Lynam see Spectacle Makers Court Minute Books, Guildhall Library, MS 5213, iv, 14. For information concerning T. S. and R. A. Cox, I am grateful to Miss S. M. Hare, Librarian of the Goldsmiths Company. The Cox association with Little Britain is commemorated by the street names Cox's Key and Albion Buildings. For Tupper see F. B. Tupper, *Family Records*, Guernsey, 1835; D. Hudson, *Martin Tupper, His Rise and Fall*, London, 1949, pp. 1–2; *Alumni Oxonienses, 1715–1856*, 4 vols., Oxford and London, 1888, iv, p. 1447. For Babington see *Burke's landed gentry*, 18th edn, 1972, 3 vols., iii, 33–4. For information that a Charles Law was a clerk at the Bank of England, I thank E. M. Kelly, Curator of the Museum and Historical Research Section, Bank of England. It has not been possible to confirm the *DNB*'s statement that Pepys was a Quaker: his name does not appear in the Digest of London and Middlesex Quarterly Meeting Births in 1775, nor in the Digest of Deaths for 1856. It is moreover stated that Pepys and two other early Askesians were not Quakers, in C. Knight, *Cyclopaedia of Biography*, London, 1858, 6 vols., iv, pp. 807–8. I wish to thank Malcolm Thomas of the Library, Religious Society of Friends, for this information. For providing references to the Aikins, I am grateful to Mrs B. Williams of the Library, Manchester College, Oxford. Dr T. D. Whittet has kindly provided information concerning apothecaries.

Table 2: *Attendance at the Mineralogical Society*

Name	Proposed	Last meeting attended	Number of meetings attended	Approximate percentage of attendance at meetings during membership
Allen, William	2 Apr. 1799 Founder	18 Dec. 1806	57	60%
Pepys, William Hasledine	2 Apr. 1799 Founder	18 Dec. 1806	73	77%
Tilloch, Alexander	2 Apr. 1799 Founder	13 Mar. 1806	35	37%
Lowry, Wilson	2 Apr. 1799 Founder	20 Nov. 1800	3	
Thezard, Alexander	16 Apr. 1799	6 Feb. 1800	2	
Lynam, Charles	16 Apr. 1799	10 Nov. 1803	16	17%
Aikin, Charles Rochement	13 June 1799	13 Mar. 1806	42	49%
Laws, Charles	5 Nov. 1799	28 May 1801	19	43%
Aikin, Arthur	3 Oct. 1799	13 Feb. 1806	44	52%
Cox, Theodor	17 Oct. 1799	24 Mar. 1803	18	50%
Phillips, Richard	6 Feb. 1800	18 Dec. 1806	51	68%
Cox, Robert Albion	6 Feb. 1800	26 Dec. 1801	7	18%
Oldham, William Chapman	17 Apr. 1800	30 Apr. 1801 (resigned 4 Feb. 1802)	17	42%
Sandman, Phillip	1 May 1800	26 Nov. 1801	17	53%
Campbell, Hector	15 May 1800	Resigned 22 Jan. 1801 and elected corresponding member	2	

Table 2 – *continued*

Name	Proposed	Last meeting attended	Number of meetings attended	Approximate percentage of attendance at meetings during membership
Tupper, Martin	21 Aug. 1800	2 Feb. 1801 and elected corresponding member	5	
Stocker, Richard	9 Oct. 1800	30 Apr. 1801	11	
Babington, William	6 Feb. 1801	4 Feb. 1802	6	
Bingley, Richard	2 Apr. 1801	18 Dec. 1806	23	51%
Cock, Thomas	13 Mar. 1806	18 Dec. 1806	3	

Notes to Table 2: Percentages of possible attendance are only given for those present at more than seven meetings.

Notes to Table 3: This is based on a summary of the Askesian Society Minutes, op. cit. (44), made perhaps with a view to recovering apparatus. These members are corroborated by Cripps, op. cit. (17), p. 27n., and Bradshaw, op. cit. (30), pp. 26–7, 47, 57. Further associates may have been Dr Relph of Guy's (d. 1804), Dr Bradley of the Westminster Hospital (d. 1813) and Thomas Poole of Nether Stowy. The summary is further confirmed by fragmentary evidence in the *Philosophical Magazine*, with the additional information that Richard Coleman of the Royal Mills, Waltham Abbey was corresponding member. Inkster, op. cit. (10), pp. 16–19 claims that G. Fordyce, A. Aikin, T. Poole and Davy were full members, and associates Tupper and Babington with Unitarianism. For Lawson's apprenticeship to Nairne, the distinguished maker of scientific instruments, see Spectacle Makers Minutes, 6 December 1788. For Quaker members, see J. Smith, op. cit. (56).

Table 3: *Profile of the Askesian Society, 1796–c. 1808*

Name	Date of election	Religion	Occupation
Allen, William	23 Mar. 1796 Founder	Quaker	Druggist and chemical manufacturer
Woods, Samuel	23 Mar. 1796 Founder	Quaker	Druggist
Mildred, Samuel	23 Mar. 1796 Founder	Quaker	Printer
Phillips, William	23 Mar. 1796 Founder	Quaker	Physician
Fox, Dr Joseph, of Finsbury	23 Mar. 1796 Founder	Quaker	Surgeon
Fox, Joseph, of Guy's	23 Mar. 1796 Founder	Quaker	
Lawson, Henry	23 Mar. 1796 Founder	Anglican (son of Dean of Battle)	Of Spectacle Makers Company
Ball, Joseph	1796		
Howard, Luke	1796	Quaker	Druggist and chemical manufacturer
Pepys, William Hasledine	26 Feb. 1799		Cutler
Tilloch, Alexander	9 Apr. 1799	became Sandemanian	Printer
Tupper, Martin	7 Oct. 1799		Medical student
Phillips, Richard	Member by 11 Feb. 1800	Quaker	Apprentice to Allen
Cooper, Astley Paston	25 Feb. 1800 (resigned 7 Apr. 1801)	Anglican	Surgeon
Babington, William	1 Dec. 1801	Anglican	Surgeon
Arch, Arthur	1 Dec. 1801	Quaker	Bookseller
Woods, Joseph	Member by 19 Feb. 1804	Quaker	Architect

that rather than representing an offensive by marginal groups, science could be used to defend social status by reinforcing established occupational, family, and propertied interests.

Acknowledgements

For help of various kinds I wish to thank Margaret Pelling, Peter Embrey, Hugh Torrens, Dr J. Squire and Irene Ashton. For permission to cite and quote from manuscripts in their keeping, I am grateful to the Royal Society; the Department of Geology and Mineralogy, University of Oxford; the Institute of Agricultural History, University of Reading; the Wills Library, Guy's Hospital; the Museum of the History of Science, University of Oxford; the Royal Institution; and the Department of Mineralogy, British Museum (Natural History).

Notes and references

1 H. B. Woodward, *The History of the Geological Society of London*, London, 1908, pp. 6–9; W. W. Watts, 'Fifty years of work of the Mineralogical Society', *The Mineralogical Magazine and Journal of the Mineralogical Society*, 1926–8, 21, 106–24 (108–9); M. J. S. Rudwick, 'The foundation of the Geological Society of London: its scheme for co-operative research and its struggle for independence', *British Journal for the History of Science*, 1963, **1**, 325–55 (326–7).

2 R. S. Porter, *The Making of Geology: Earth Science in Britain 1660–1815*, Cambridge, 1977, p. 147.

3 R. S. Porter, 'The Industrial Revolution and the rise of the science of geology', in M. Teich and R. Young (eds.), *Changing Perspectives in the History of Science: Essays in Honour of Joseph Needham*, London, 1973, p. 325. For an alternative view: M. Guntau, 'The emergence of geology as a scientific discipline', *History of Science*, 1978, **16**, 280–90.

4 A. E. Musson and E. Robinson, *Science and Technology in the Industrial Revolution*, Manchester, 1969, p. 138. For criticism see: J. R. Harris, 'Skills, coal and British industry in the eighteenth century', *History*, 1976, **61**, 167–82.

5 P. Mathias, 'Who unbound Prometheus? Science and technical change, 1600–1800', in A. E. Musson (ed.), *Science, Technology and Economic Growth in the Eighteenth Century*, London, 1972, p. 91, stresses the importance of new attitudes.

6   D. T. Andrew, 'London charity in the eighteenth century', Ph.D thesis, University of Toronto, 1977, p. 283.

7   M. D. George, *London Life in the Eighteenth Century*, London, 1925, p. 11; M. B. Buer, *Health, Wealth and Population in the Early Days of the Industrial Revolution*, London, 1926; J. L. Hammond and B. Hammond, *The Town Labourer*, London, 1917.

8   A. S. Collins, *The Profession of Letters: A Study of the Relation of Author to Patron, Publisher and Public, 1780–1832*, London, 1928; J. Summerson, *Georgian London*, London, 1945; B. Lillywhite, *London Coffee Houses: A Reference Book of Coffee Houses of the Seventeenth, Eighteenth and Nineteenth Centuries*, London, 1963; and W. B. Todd, *A Directory of Printers and Others in Allied Trades: London and Vicinity, 1800–1840*, London, 1972, have made little impact on general historians. An exception is G. Rudé, *Hanoverian London, 1714–1808*, London, 1971. For science and politics in London, see P. Weindling, 'Science and sedition: how effective were the Acts licensing lectures and meetings, 1795–1819?', *British Journal for the History of Science*, 1980, **13**, 139–53; I. Inkster, 'London science and the Seditious Meetings Act of 1817', ibid., 1979, **12**, 192–6; and Inkster, 'Seditious science: a reply to Paul Weindling', ibid., 1981, **14**, 181–7.

9   M. Berman, *Social Change and Scientific Organization: The Royal Institution, 1799–1844*, London, 1978, p. 75.

10 I. Inkster, 'Science and society in the metropolis: a preliminary examination of the social and institutional context of the Askesian Society of London, 1796–1807', *Annals of Science*, 1977, **34**, 1–32 (3–4, 16–27).

11 Minute book of the Mineralogical Society, Department of Mineralogy, British Museum (Natural History).

12 MS Gunther 4, Museum of the History of Science, Oxford, 28 November 1783, 23 January 1784, 15 April 1795, 30 April 1795.

13 B. Higgins (ed.), *Minutes of the Society for Philosophical Experiments and Conversations*, London, 1795; F. W. Gibbs, 'Bryan Higgins and his circle' in Musson, op. cit. (5), 195–207.

14 Royal Society MS, 155.

15 W. Babington, *MS Lectures on Chemistry*, n.d., pp. 1–276 (2–3), Guy's Hospital Library.

16 R. W. Horne, 'Pharmacy at Guy's 1726–1976', *Pharmeceutical Journal*, 1976, **218**, 395–7 (395). For further associations of Mineralogical Society members with the Chemical and Physical Societies of Guy's Hospital, see: Minutes of the Physical Society, Guy's Hospital library; Allen was proposed 3 October 1795, Knight on 27 October

1798, Davy on 27 December 1800; Babington, Fox (of Guy's), Stocker, A. Cooper and possibly Tupper were members.

17 Musson and Robinson, op. cit. (4), p. 138; E. C. Cripps, *Plough Court, the story of a Notable Pharmacy, 1715–1927*, London, 1927, pp. 25–53.

18 A. Clow and N. L. Clow, *The Chemical Revolution: A Contribution to Social Technology*, London, 1952, pp. 232–3.

19 'Account of schools for public service, in the French Republic', *Monthly Magazine*, 1799, **7**, 114–18 (116–17). I thank Hugh Torrens for this source. For further notices of the BMS see: *The Royal Kalender, or, Complete and Correct Annual Register for England, Scotland, Ireland and Wales*, London, 1803, p. 281; J. Feltham, *The Picture of London for 1802*, London, 1802, p. 182; *Philosophical Magazine*, 1799, **3**, 318; 1801, **6**, 1–3, 369–72; 1801–2, **9**, 282–3; 1802–3, **12**, 284–7; 1804, **19**, 85–94; *Nicolson's Journal of Philosophy*, 1799, **3**, 138–41.

20 Minutes, op. cit. (11), 6 January 1803.

21 Porter, op. cit. (2), p. 140.

22 R. Burt, *John Taylor, Mining Entrepreneur and Engineer*, Buxton, 1977.

23 F. M. Osborn, *The Story of the Mushets*, London, 1951, p. 30.

24 C. Phillips, *A Short Memoir of William Phillips*, London, 1891.

25 R. Warner, *A Tour through Cornwall in the Autumn of 1808*, London, 1809, p. 241; D. B. Barton, *Essays in Cornish Mining History*, Truro, 1971, **ii**, p. 15. I thank Peter Embrey for references to Taylor and Williams.

26 Farey to Smith, 1 October 1802, Smith Collection, Department of Geology and Mineralogy, University of Oxford. Smith also contacted Tilloch presumably for advice on publishing.

27 Farey to Smith, 21 February 1805, Smith Collection.

28 *Monthly Magazine*, 1803, **16**, 4, 103, 218, 232, 380.

29 A. Aikin, *Journal of a Tour through North Wales and Part of Shropshire with Observations in Mineralogy and Other Branches of Natural History*, London, 1797, pp. v–vi.

30 For tours of Allen, J. Woods and S. Woods, L. Howard and A. Aikin, see: L. W. Dillwyn and D. Turner, *The Botanist's Guide through England and Wales*, London, 1805; L. Bradshaw (ed.), *Life of William Allen with Selections from his Correspondence*, 3 vols., London, 1846–7, **i**, pp. 23–5; L. Howard, MS Diary of a tour to the West Riding and Cumberland in 1807, Eliot and Howard Papers,

Greater London Record Office (Middlesex Records), Acc. 1017/ 1397.

31 M. W. Flinn (ed.), *Swedenstierna's Tour in Great Britain 1802–3: The Travel Diary of an Industrial Spy*, Newton Abbot, 1973, pp. 16, 19–20. I thank Hugh Torrens for this reference.

32 G. B. Greenough, MS Diary of a tour to Scotland in 1805, Greenough Papers, University College London, 15 August 1805.

33 T. E. Thorpe, *Essays in Historical Chemistry*, 3rd edn, London, 1911, p. 559; L. J. Spencer, 'The "satin spar" of Alston in Cumberland; and the determination of massive and fibrous calcites and aragonites', *The Mineralogical Magazine and Journal of the Mineralogical Society*, 1895–7, **11**, 184–7.

34 Porter, op. cit. (2), p. 146.

35 *Philosophical Magazine*, 1802, **13**, 153–5, 253–7; 1803, **17**, 166–8, 253–9.

36 *Monthly Magazine*, 1801, p. 358.

37 Minute books of the Board of Agriculture, Institute of Agricultural History, 16 March 1802, 27 May 1803, 30 May 1803, 30 April 1805.

38 W. Babington, op. cit. (15), p. 58.

39 A. Tilloch, 'A new process for rendering platina malleable', *Philosophical Magazine*, 1805, **21**, 175; D. McDonald, *A History of Platinum*, London, 1960, pp. 83–119; Thorpe, op. cit. (33), p. 559; on Johnson see *Journal of the Chemical Society of London*, 1867, **20**, 392–5; Banks to Pepys, 30 May 1805 and 16 June 1805, Pepys Papers, Royal Institution. P. J. Weindling, 'A platinum gift to King George III: a gesture by William Hasledine Pepys, cutler and instrument maker', *Platinum Metals Review*, 1982, 26, 34–7

40 D. M. Knight, 'The vital flame', *Ambix*, 1976, **23**, 5–15 (10).

41 Berman, op. cit. (9), pp. xxiv–xxv, 51–2, 70–4.

42 *Journal of the Royal Institution* , 1802, **1**, 66, 84.

43 P. J. Weindling, 'Geological controversy and its historiography: the prehistory of the Geological Society of London', in L. Jordanova and R. S. Porter (eds.), *Images of the Earth: New Essays in the History of the Environmental Sciences*, Chalfont St Giles, 1979, 248–71; *Monthly Magazine*, 1804, **16**, 57, 564.

44 Summary of Askesian Society minutes, Eliot and Howard Papers, op. cit. (30), Acc. 1017/1492.

45 G. B. Greenough, MS History of the Geological Society of London, Greenough Papers, Cambridge University Library; Inkster, op. cit. (10), pp. 27–30.

46   London Institution Papers, Guildhall Library, MS 3080.
47   J. Cutler, 'The London Institution 1805–1933', University of Leicester Ph.D thesis, 1976, pp. 8, 15, 86, appendix; L. C. Ockenden, 'The great batteries of the London Institution', *Annals of Science*, 1937, **2**, 183–4.
48   Eliot and Howard Papers, op. cit. (30), Acc. 1017/1491.
49   A. Aikin, *A manual of Mineralogy*, London, 1814, p. v; W. Phillips, *An Elementary Introduction to Mineralogy*, London, 1816, pp. i–v.
50   Berman, op. cit. (9), p. 75.
51   Inkster, op. cit. (10), p. 14.
52   Rudé, op. cit. (8), p. xi; W. D. Rubinstein, 'Wealth, elites and the class structure of modern Britain', *Past and Present*, 1977, no. 76, 99–126 (104–13); N. Rogers, 'Money, land and lineage: the big bourgeoisie of Hanoverian London', *Social History*, 1979, **4**, 437–54 (442–4); L. D. Schwarz, 'Income distribution and social structure in London in the late eighteenth century', *Economic History Review*, 1979, **32**, 250–9.
53   Inkster, op. cit. (10), p. 4.
54   E. Isichei, *Victorian Quakers*, Oxford, 1970, p. 177; J. Nicholl, *Some Account of the Worshipful Company of Ironmasters*, 2nd edn, London, 1866, pp. 347–8, 609–10; Inkster, op. cit. (10), p. 31.
55   A. W. Slater (ed.), 'Autobiographical memoir of Joseph Jewell 1763–1846', *Camden Miscellany*, 1964, **22**, 113–78 (123).
56   Joseph Smith, *A Descriptive Catalogue of Friends' Books*, 2 vols., London, 1867, **ii**, 736–7; D. Richards, 'Destiny or dynasty: the Fox family of doctors and dentists – a seven-generational contribution to the British health services', *Bulletin of the Society for the Social History of Medicine*, 1975, **16**, 11–13; S. Wilks and G. T. Bettany, *Biographical Dictionary of Guy's Hospital*, London, 1892, p. 199.
57   C. Welch, *History of the Cutlers Company of London*, 2 vols., London, 1923, **ii**, pp. 274, 280; Phillips, op. cit. (24); J. Z. Fullmer, 'Davy's sketches of his contemporaries', *Chymia*, 1967, **12**, 127–50 (135); W. C. Pepys, *Genealogy of the Pepys Family 1273–1887*, London, 1887. I thank Robert Bud for this reference.
58   M. J. Peterson, *The Medical Profession in Mid-Victorian London*, Berkeley and London, 1978, pp. 146–51, 291–4.
59   Wilks and Bettany, op. cit. (56), pp. 235–8, 410–13, 434–9. Burke's *Landed Gentry of Ireland*, 4th edn, London, 1958, provides for Babington some confirmation. I thank Hugh Torrens for this suggestion.

60   The Askesians also spent large amounts on apparatus. See Eliot and Howard Papers, op. cit. (30), Acc. 1017/1491.

61   J. H. M. Craig, *A History of the London Mint from AD 287 to 1948*, Cambridge, 1953, pp. 259, 302.

62   Welch, op. cit. (57), ii. pp. 243–4.

63   R. Phillips, *An Experimental Examination of the Last Edition of the Pharmacopoeia Londoniensis, with Remarks on Dr Powell's Translation and Annotations*, London, 1811; T. D. Whittet, 'Clerks, bedels and chemical operators of the Society of Apothecaries. The Gideon de Laune Lecture for 1977' (typescript), fos. 29–31.

64   Berman, op. cit. (9), pp. 103–4.

65   Wilks and Bettany, op. cit. (56), pp. 472–3.

66   R. S. Porter, 'Gentlemen and geology: the emergence of a scientific career, 1660–1920', *Historical Journal*, 1978, **21**, 809–36 (816–17).

67   Allen's certificate (Royal Society Certificates, vi, 121) was signed by E. Howard, Babington, A. Cooper, Dillwyn, Hatchett, Englefield, Guillemarde, Bourbon and Dawson Turner.

68   L. Aikin, *Memoir of Dr John Aikin*, 2 vols., 1815, i, 247–8; *Christian reformer*, 1847, **3**, 323–32 (330), and 1854, **10**, 379–80. For Tilloch see: *Mechanic's Oracle*, 1825, **1**, 220; *Philosophical Magazine*, 1825, **65**, 134–5; *Gentleman's Magazine*, 1825, **95**, 276–81; *Imperial Magazine*, 1825, **7**, 208–22; *Annual Biography and Obituary*, 1826, **10**, 320–34 (333). Lowry was said to have 'preferred the theory of the republican and the practice of the aristocracy': *Annual Biography and Obituary*, 1825, **9**, 93–107 (103).

69   Thorpe, op. cit. (33), p. 565; Weindling, op. cit. (43), p. 265.

70   Inkster, op. cit. (10), pp. 25–7.

71   Bradshaw, op. cit. (30), i, pp. 33, 36; MS Minutes of the committee, Greater London Record Office P98/CTC1/55; 'An account of the Soup Society in Spitalfields', *The Philanthropist*, 1812, **2**, 173–200; P. McCann, 'Popular education, socialization and social control: Spitalfields 1812–1824', in P. McCann (ed.), *Popular Education and Socialization in the Nineteenth Century*, London, 1977, 1–42 (12–18).

72   B. K. Gray, *A History of English Philanthropy*, London, 1905, pp. 275–90; B. Thompson, 'Of food; and particularly of feeding the poor' (1795), in S. C. Brown (ed.), *Collected Works of Count Rumford*, 5 vols., Cambridge, Mass., v, pp. 169–262 (172); Isichei, op. cit. (54), p. 177.

73   J. F. C. Harrison, *Robert Owen and the Owenites in Britain and America*, London, 1969, pp. 20, 223; A. Prochaska, 'The practice of

radicalism: educational reform in Westminster', in J. Stevenson (ed.), *London in the Age of Reform*, Oxford, 1976, pp. 102–16 (112).

74   J. A. Hone, 'Radicalism in London 1796–1802: Convergeances and continuities', in Stevenson, op. cit. (73), pp. 79–101 (84, 94, 101); B. B. Cooper, *The Life of Sir Astley Cooper*, 2 vols., London, 1843, **ii**, pp. 294–8.

5 'Nibbling at the teats of science': Edinburgh and the diffusion of science in the 1830s

Steven Shapin

The diffusion of science is often conceived of as a passive process, as the osmotic transfer of ideas from regions of high "truth concentration" into areas of low concentration. Such a conception has little explanatory potential. It is more historically rewarding to recognize the diffusion of science as an active process, undertaken by specifiable social groups for their particular purposes, in contexts which provide both resources and constraints for any given enterprise of diffusion.[1] In other words, the diffusion of scientific knowledge should be seen as a political and a logistic problem. The boundaries across which knowledge is diffused may vary: between countries, between town and country, between social classes, and so on. Movement across each sort of boundary poses its special political and logistic problems, and the contexts in which the movement takes place add their particularities to the problems. These include the interests of social groups in transferring knowledge, the reaction of other groups to the purposes of those initiating the transfer, and the means available to effect the ends. In each instance, the political and logistic problems reflect back upon the knowledge it is proposed to transfer. Why *this* sort of knowledge and not *that*? The success or failure of diffusion hinges on contextual perceptions of science as a potential resource; its instrumental potential is assessed both by groups engaged in facilitating its diffusion and by groups on whose approval or co-operation the successful diffusion depends.

Although these programmatic recommendations about how to approach the diffusion of science may be generally applicable, their value must be assessed by demonstrating their pertinence to the history of particular local enterprises. For that reason this essay is concerned with specific plans for the diffusion of science which were developed in Edinburgh during the 1830s. That one the enterprises did not even come into being is immaterial: failure often exhibits more clearly than success the constraints and resources available for the diffusion of science.

The Edinburgh context

The paradox of Edinburgh in the early nineteenth century was that it was a provincial metropolis. No longer the capital of an independent Scotland, its legal, ecclesiastical and financial institutions nevertheless exerted considerable control of "North British" affairs. Stripped of its Parliament, Edinburgh danced to Westminster's tune, while the Scottish hinterlands danced to Edinburgh's. However, in cultural matters Edinburgh still functioned as the metropolis of a vast empire of science. Edinburgh-trained savants pressed out beyond the Scottish provinces to staff the medical and technical services of the British Empire, to found the new scientifically orientated University College of London, and to instigate the establishment of several English 'literary and philosophical' societies and 'Mechanics' Institutes'. The itinerating paths of scores of scientific lecturers traced back to Edinburgh classrooms.[2]

Nor did Edinburgh's vast export market diminish the ferment of science that brewed within the city itself. A plethora of scientific, medical and ·technical organizations catered for local men of science, from the prestigious Royal Society of Edinburgh to the iconoclastic Phrenological Society, from the utilitarian Scottish Society of Arts to the bibulous Aesculapian Society. The *Edinburgh Review* assessed science elegantly and whiggishly; *Blackwood's* Tory ripostes manifested more wit than weight. Edinburgh's then flourishing publishing and instrument-making industries supplied the wants of scientific sciolist and specialist. Surveying the scene in 1826 the visiting John James Audubon judged Edinburgh to be 'the very vitals of science'.[3]

Yet against this background of apparent vitality many local commentators sounded stylized laments on the theme of the *decline* of Edinburgh science from what it had been during the greatest years of Enlightenment production in the late eighteenth century. Edinburgh was seen in crepuscular transit from having been 'the Athens of the North' to becoming 'the Reykjavik of the South'.[4] What was in fact occurring in the 1820s and 1830s was a change in the boundaries of participation in scientific culture, a change which many observers either regretted or under-valued. A hegemonic old order was passing away. Previously 'too subservient to be feared', newly self-confident and increasingly less deferential mercantile and commercial groups were finding their cultural voices at the same time as they were finding their political voices.[5] The great Whig lawyer Henry Cockburn was one who generally approved of this development, although not without a hint of condescension:

The intellectual fermentation is astonishing. On the single subject of popular instruction, any newspaper states facts and suggests reflections which imply the dawn of a new day. Schools, lectures, private colleges, normal institutions for the mere manufacture of teachers, pedagogue mills, associations open to all ranks, but chiefly for the middle and lower, inviting both sexes and all ages, embracing all subjects, physical and moral, practical and theoretical. . . . There is some quackery and a great deal of superficial absurdity in all this. . . .[6]

The Edinburgh Whigs initially aimed "benevolently" to assist and control the new enterprises. Leonard Horner established the School of Arts in 1821, for the scientific instruction of artisans, successfully stipulating that its curriculum and management should remain in the hands of "the higher ranks".[7] The School prospered, prompting Horner's fellow Edinburgh Whig Henry Brougham to launch the great 1825 campaign to establish Mechanics' Institutes throughout Great Britain. However, the scientific "fermentation" among Edinburgh's mercantile groups soon passed beyond the capacity of either Whigs or Tories of "the higher ranks" to control. Certain mercantile classes began to help themselves to scientific culture.

Petty-bourgeois science in 1830s Edinburgh

In the same year that the Reform Bill swept aside much of Tory "Old Corruption",[8] Henry Cockburn remarked upon the founding of a novel sort of scientific organization in Edinburgh:

An establishment, called 'The Edinburgh Association for procuring Instruction in Useful and Entertaining Science', was set up here in 1832. . . . This and similar institutions are strongly characteristic of the times. It is a sort of popular unendowed college, where lectures are given to all, male or female. . . . The lectures are on botany, geology, chemistry, astronomy, physiology, natural philosophy, phrenology, and education. . . . It is a very useful establishment, giving respectable discourses very cheaply to a class of persons for whose scientific instruction and amusement there is no other provision. . . . They are of course contumelious of colleges, and are rather more conceited of their knowledge than humble of their ignorance. George Combe is their genius, and consequently phrenology is a favourite and most productive branch. The poor classics are held in utter scorn. In spite of these follies it is gratifying to see hundreds of clerks and shopkeepers, with their wives and daughters, nibbling at the teats of science anyhow.[9]

This society, later called the Edinburgh Philosophical Association (henceforth EPA), represented the first intrusion into organized literate culture by the Edinburgh "petty bourgeoisie". If the social contours of British mercantile groups in the early nineteenth century are normally difficult to retrieve, those of Edinburgh are even more so. Edinburgh was not a significant centre of industrial manufacture or colonial trade: it did not produce a powerful class of "great merchants" such as that thrown up by Manchester or Glasgow in the late eighteenth century. In contrast the Scottish metropolis was a hive of small workshops crafting the articles demanded by Edinburgh's professionals and aristocrats: books and prints, jewellery, quality clothing, furnishings, and the like. Large numbers of "black-coated" clerks served its counting-houses and bigger shops. It was the clerk, the smaller shopkeeper, and the self-employed craftsman who comprised the local petty-bourgeoisie – in general, a group acutely conscious of its "respectability", equally strongly insistent on its difference from the working classes "proper", determinedly upwardly mobile, yet permeated by a "fear of falling".[10] It was the petty-bourgeoisie which founded the EPA and gave it its original aims and form. The historical significance of the EPA lies in the way it manifested the politics of science in a provincial metropolis, and the manner in which it brought together the interests of three elements in local scientific culture: the "clerks and shopkeepers", the scientific lecturers, and the Edinburgh phrenological community around George Combe: interests which were balanced only with great difficulty.

The most original feature of the EPA was that it was controlled by those social groups which the organization was designed to instruct and amuse: its management was in the hands of its mercantile membership. The social standing of that membership was precisely defined and advertised; they claimed to be 'respectable tradesmen' and 'generally above the class who work for days' wages'.[11] The first year's Directors thus included a typical roll-call of Edinburgh's self-employed master-artisans and shopkeepers: two engravers, two jewellers, two booksellers, two clothiers, and one architect, builder, tailor, die-cutter, hatter, engineer, seedsman, hat-maker, linen-draper, and boot-maker.[12] 'There are among them', it was boasted, 'no influential literary or scientific characters.'[13] And the members of the Association proudly called themselves 'merchants in philosophy – barterers of thought – traders in knowledge'.[14]

The enthusiastic middle-class newspaper *The Scotsman* approved the EPA's defiant independence from established cultural institutions:

It is a proud boast for Edinburgh, that *there*, for the first time in the history of society, the commercial and business classes, for whom the gates of Colleges have not been wont to lift up their heads, have, in these brighter days for mankind, 'risen', and demanded science for themselves. . .and without patronage, nay, in the face of some sneers and discouragements, achieved for themselves an easy path to the temple of scientific light.[15]

Edinburgh's "mechanics", having been provided for by Horner's School of Arts, were now joined in science by 'our shopkeepers and clerks' who naturally insisted on a separate and more elevated institution.[16] 'The higher orders', *The Scotsman* warned, 'have, in truth, need to bestir themselves, if they do not wish to be left behind in the march of improvement.'[17]

The lectures were cheap, as befitted an Association which had condemned the 'high fees payable to the professors in the University'.[18] One could have a course of approximately twenty-five lectures in geology for 7s. 6d.; chemistry for 10s. 6d.; phrenology and physiology for 10s. 6d.; or the lot for £1. And the lectures were conveniently timed, after working hours, at half-past eight in the evening, a time made more acceptable by improvements in gas lighting. Furthermore, it was intended that the Association's members should have the subjects they wished, taught by the lecturers who pleased them best. To the early range of subjects were soon added botany, education, natural philosophy, astronomy, zoology, and animal economy. By the late 1830s the EPA's curriculum had expanded to take in natural history, 'electricity, galvanism and magnetism', as well as non-scientific subjects like jurisprudence, oratory, constitutional history, drama, and vocal harmony.

From its founding until 1836 the EPA showed a marked preference for the natural sciences, the study of which was justified by their social and moral consequences. The sciences as a whole were well adapted to diffusion through the social strata as they were 'surely capable of being explained in language intelligible to every ordinary mind', because they 'elevated the mind' and quickened the sense of wonder at the Creator's works, and because they acted as an antidote to improper and unworthy thoughts: 'Science is not only valuable for what it makes us do, but for what it prevents us from doing.'[19] Botany was favoured as 'especially adapted to the female mind' and because 'the exquisite harmony of design' which its study reveals 'cannot fail to afford the highest gratification';[20] physiology was recommended as the means of preserving health, and removing 'temptations to moral depravity', especially when disseminated to 'the indigent and industrious classes'.[21] The Association was hugely successful. Within a year of its founding, its lectures were reaching several thousand of Edinburgh's mercantile classes. The EPA's meticulous accounts show its receipts approached £1000 and its profits ran to several hundred pounds. Petty-bourgeois science was good business, and Edinburgh's shopkeepers were well pleased with their foray into the merchandising of scientific culture.

What of the scientific lecturers' interests in the EPA? In the highly competitive environment of the upper echelons of local scientific affairs, lecturers also had reason to be pleased that another forum for remunerative performances had been established. While the EPA's Directors were not profligate, neither were their fees contemptible. Typical sums paid for courses of twenty-five to thirty lectures were the fifty guineas to Dr John Murray for a geology and chemistry series, £40 to Dr James L. Drummond for botany, and precisely £91. 7s. 4d. to George Combe for a phrenology course. Invitations to lecture at the Association were eagerly sought; and, given the embarrassing wealth of scientific lecturing talent available in Edinburgh, there was no lack of choice for the Directors of the EPA. It was a buyers' market for scientific lectures, and the EPA was committed to treating its lecturers as its mercantile members were used to treating their employees. The emergent Edinburgh petty-bourgeoisie did not propose to be dictated to by intellectuals, just as they did not mean any longer to be patronized by lords and lawyers. Sound commercial sense told them that who pays the piper calls the tune. Lecturers were appointed for only one session at a time; subjects were selected according to interests revealed by a canvass of members.

If any lecturer fails to interest and instruct his audience, he will have small chance of being selected in a succeeding year. . . . [The Directors will] be decided and uncompromising in rejecting every lecturer who had not given ample satisfaction to his audience. They ought to allow no feelings of private friendship or supposed delicacy to individuals to induce them to tolerate feebleness, ignorance, or want of talent for instructing.[22]

It was to be survival of the lecturing fittest, and the 'merchants in philosophy' were stern masters.

The third force involved in the EPA's affairs was George Combe and the Edinburgh phrenological community.[23] Accounts of the founding of the EPA differ slightly, but all agree in tracing it back to Combe's popular lectures on phrenology which he began offering in Edinburgh in 1825 to audiences of over 100.[24] After the controversy surrounding the publication of Combe's *The Constitution of Man* in 1828, local interest in phrenology (or, at least, willingness publicly to be associated with it) declined, and the lectures were suspended in 1831–2.[25] Agitation over Reform, and perhaps Combe's own position as Secretary of the Edinburgh Reform Committee resulted in a changed climate, and his popular course was resumed during May–June 1832. At the end of this course, several petty-bourgeois members of the audience, apparently led by William Fraser, the printer of the small newspaper *The Edinburgh Weekly Chronicle*, petitioned Combe 'for a more extended course during winter, along with Lectures on some other subjects of Natural Science'.[26] These lectures (Combe on phrenology, Murray on geology and chemistry) duly began in November 1832, under the auspices of the new Association.

If Edinburgh was the metropolis of an empire of science, George Combe was one of its leading imperialists. After his conversion to Gall's and Spurzheim's cranioscopic doctrines in 1815, the Edinburgh lawyer and son of a small brewer developed ambitious plans to diffuse not just phrenology but also fundamental knowledge of the natural sciences through the social strata and across Great Britain.[27] His horizons, unlike those of the Edinburgh petty-bourgeoisie, were national and even international. Wide popular knowledge of "the natural laws" governing man's constitution was his major aim. With the founding in 1820 of the Edinburgh Phrenological Society and *The Phrenological Journal* in 1823, he gathered around him a coterie of followers equally committed to the diffusion of science in the cause of social reform. Rebuffed by Leonard Horner when he offered to give a course in phrenology for the Edinburgh School of Arts,[28] Combe nevertheless carried the message around scores of provincial Mechanics' Institutes and literary and philosophical societies. In 1835 Combe and fellow phrenologists Sidney Smith, W. B. Hodgson and James Simpson founded the Edinburgh Society for the Diffusion of Moral and Economical Knowledge, which offered the working classes penny lectures on subjects proscribed by the narrowly scientific curriculum of the School of Arts.[29] In 1836 this was succeeded by twopenny phrenological, chemical and political-economical lectures given at the Freemasons' Hall by the Association of the Working Classes, for their Social, Intellectual, and Moral Improvement, directly modelled on the EPA, but adapted for the labouring classes.[30] To supply and to create demand for his works in these and similar institutions, Combe used the funds provided by the Henderson Bequest to print tens of thousands of copies of *The Constitution of Man* to be sold for 1s. 6d.[31] Periodic attempts were also made to diffuse phrenology upwards to the intellectual elite: in the mid 1830s Hodgson pressed for a chair of phrenology to be established in the University of Edinburgh[32] and in 1836 Combe himself unsuccessfully contested the chair of logic.[33] In 1838 Combe collaborated in the institution of the Phrenological Association, an itinerant body which initially followed the British Association for the Advancement of Science around the country and then collapsed in 1842.[34] All this cultural imperialism, it needs to be repeated, was not confined to the spread of cranioscopy: Combe's phrenology was the most important vehicle for the diffusion of naturalistic and materialistic views in early to mid nineteenth-century Britain.[35] The diffusion of physiology, anatomy, geology, astronomy, and chemistry across class boundaries in the period cannot be understood without giving serious attention to the phrenological vehicle and the social programmes phrenologists intended to further.

Combe's secularizing and reforming programmes aimed to unite a broad range of bourgeois interests. The diffusion of phrenological and naturalistic culture to social groups arrayed below was intended to forge effective alliances with them against reactionary and clerical forces which stood opposed to 'improving' and 'rational' social and cultural change. Not surprisingly, the Combeites encountered severe resistance from Edinburgh Tories and Whigs: the former because they were hostile to Combe's favoured reforms, and the latter largely because they wished themselves to control the course of reform and to frame alliances with lower-class groups.

Sectors of the Presbyterian Kirk attacked Combe's doctrines as irreligious; custodians of classical education assailed the 'indelicacy' of physiological lectures to a 'promiscuous' audience, the 'barbarism' of his cavalier attitude towards the value of Greek and Latin, and pointed to phrenology 'as the disturber of the peace and well-being of society'.[36]

From its inception in 1832 Combe played an active role in the affairs of the Edinburgh Philosophical Association. But that role was a difficult one. Although he was the 'genius' of the EPA, as Cockburn recognized, Combe was nevertheless the non-tenured employee of the very petty-bourgeoisie whose enlightenment and support he sought. He had ambitions for the EPA, just as he had for the diffusion of scientific culture throughout Britain. But Combe had to tread carefully; the Edinburgh petty-bourgeoisie were wary of outside leaders, even those with whose views they had such basic sympathy. For three years until the autumn of 1835, the EPA successfully harnessed the interests of the three elements which it comprised: the Edinburgh petty-bourgeoisie, the scientific lecturers, and the Combeites. This particular exercise in the transfer of knowledge from scientist to shopkeeper depended upon harmonizing social interests which, at any time, could emerge as incompatible. Towards the end of 1835 the Combeites, the scientific lecturers, and new allies developed a plan for the diffusion of science which disrupted existing accord.

Edinburgh science for the provinces

In September 1834 the three-year old British Association for the Advancement of Science henceforth (BAAS) met in Edinburgh partly under the auspices of the Royal Society of Edinburgh and its General Secretary John (later Sir John) Robison. It took quite a show to stir up Edinburgh in September, but BAAS managed the feat. Even Henry Cockburn was impressed:

I thought this a useless institution until I saw it. But I was wrong. It does good in three ways – 1. the mere gathering of philosophers . . . is exciting and convenient, and promotes their future intercourse; 2. the discussions in the sections are always upon deep and important matter . . . 3. the evening proceedings brought a crowd together, which, if not instructed, was at least amused . . . and this had a tendency to diffuse a growing taste for such subjects. The worst of it is the affectation of science in the tail.[37]

The best of BAAS, according to Cockburn, was Arago's scientific demonstration 'that, however incredible it might appear, the sun *must* shine sometimes in Edinburgh'.[38]

As ever the outsider, Combe's offer to display his skull collection before BAAS was declined, as was his later request to become a member of BAAS's General Committee.[39] Nevertheless, Combe attended the Edinburgh BAAS, as did many of his friends in the scientific lecturing business. An idea was planted. In spring 1835 Combe and Robison were in frequent contact. On 24 October 1835 a public dinner was held in honour of the Reverend J. P. Nichol whom Combe had recommended to the EPA and other scientific organizations as a lecturer in astronomy and geology.[40] The dinner was attended by such Combeites as the advocate and educationalist James Simpson, the publishers Robert and William Chambers, the printer William Fraser (a Director of the EPA) and the phrenologist-writer Sidney Smith. Three days later Robison suggested that Combe should meet him and Nichol to discuss the idea further, and the three men duly met at Robison's house on 4 November.[41] The notion, which originated with Nichol, was to form an agency for supplying scientific lecturing talent from Edinburgh to the host of provincial societies represented at BAAS.[42] Combe immediately communicated the idea to some of his influential friends, in particular to the phrenologist-baronet Sir George Stewart Mackenzie:

All over the country [Combe wrote] Associations are forming for obtaining Lectures on useful science, but there are no Lecturers, and no means of procuring them. Unless some means be adopted to direct this movement, it will either die away, or become erratic. Mr. Nichol therefore proposes that in Edin[r] we should form an Association for the purpose of directing it, & assisting its beneficial action, similar to the Society for the promotion of useful knowledge in London [the SDUK]. This society might correspond with local Associations, become the medium of finding out and certifying the capacity of Lecturers, recommending books &c.[43]

Nichol, whose scientific capacities had hitherto found outlets primarily in school-teaching positions in the Scottish provinces, thought that such an association might be useful both to himself and to the Combeite reforming cause. By giving 'sample lecture courses in each [provincial] town ... I might stir up a commotion in many districts'. 'In the course of two seasons at the utmost,' he wrote to Combe, 'the greater part of Scotland might be gone over; & surely we shall be able in some way to breed up a few young men with faculties fitted to keep the fire awake.'[44]

Nichol's conception of the organization may also help to explicate the role of Robison whose interest is not made clear in any documentary evidence. Early in 1835 Nichol had suggested to Combe 'the consolidation of an Institution for *practical* Science', which might be useful in training civil engineers.[45] Robison, as well as being General Secretary of the Edinburgh Royal Society, was an active engineer and mechanical inventor.[46] It is possible that both Nichol and Robison hoped that the association, once set on foot, might further engineering interests.

By 9 January 1836 Combe was able to announce in *The Scotsman* and *The Phrenological Journal* the formation of the Society for Aiding the General Diffusion of Science. The initial notice for the new Society pointed out that the Edinburgh Philosophical Association had provided the model for 'the institution of a number of provincial associations, essentially similar to that of Edinburgh in their constituent elements and objects'.[47] From these provincial associations (in Dalkeith, Kirkcaldy, Montrose, and elsewhere) the EPA had apparently received a large number of requests 'for advice and assistance in the general direction of their efforts, and especially in procuring, through them, the services of able and qualified lecturers'. Combe was confident that 'the good which may be accomplished by means of scientific lectures delivered to popular audiences in a luminous style, with the requisite illustrations, is incalculable; and any means which can be adopted to advance it, must contribute to the ultimate advantage of the nation'. The plan was to form in Edinburgh, a 'central society', a kind of clearing-house to service the provinces by sending out 'qualified lecturers', which *de facto* meant lecturers qualified to preach the Combeite message of scientific culture and social improvement.

This remarkable plan presented considerable political and logistic problems. The most intractable one, as Combe acknowledged to Mackenzie, was that the new Society was likely to be perceived as the ideological device the phrenologists designed it to be. How could it retain its goals while constituting its membership on a broad enough base to carry the aims into effect?

The difficulty is to determine who should be members. If we admit Tories, they are leagued with the Church, and the clergy hate us and are preaching against us. If we appeal to the Whigs, they do not like us as Phrenologists, and the infusion of a number of their names would give the society a political cast. We propose therefore to confine it only to men of liberal sentiments . . . and to consult the Whig leaders to ascertain whether they do not object; but not to ask their co-operation.[48]

Accordingly, on 14 December 1835 Combe wrote to William Murray, a rich Edinburgh gentleman whose brother John Archibald Murray had just become Whig Lord Advocate for Scotland.[49] As a result, Combe said, of the success of the EPA and 'of the wide diffusion of my work on the constitution of man among the operative classes, there is a general & earnest call over Scotland for lecturers on science to instruct the people'. The object of the proposed Society, he told Murray, would be 'to teach the people science & sound morality, but resolutely to steer clear of religion, leaving to the spiritual guides of the people the duty of supplying that instruction, each according to his own views'. The doctrines of *The Constitution of Man*, Combe freely admitted, would be 'the leading principles on which [the Society] would probably act', and, on that basis he solicited the co-operation of the leading Edinburgh Whigs. Meantime, Mackenzie, whose phrenological sympathies were not in doubt but whose politics were suspect, needed further reassurance. We do not, Combe told him, intend to send out 'missionaries to stir up the people'; '[w]e wish to give the Society so humble an aspect that it may not tempt the fanatical & Tories to establish a counter society to obstruct our progress'; '[i]n fact, our object will be to encourage and instruct *lecturers*, and not to excite the people, for they already demand more instruction than we can supply'.[50]

Once again Combe approached Leonard Horner of the School of Arts and once again he was rebuffed, Horner mistakenly assuming that he was being asked to become a Director of the EPA and not of the new Society for Diffusion.[51] The great Whig educationalist objected to the organizational structure of the EPA; it is an 'error in principle', Horner pronounced, 'that in such a place of instruction as the Philosophical Association the exclusive management should be vested in the persons to be instructed. . . .I am moreover of opinion that it is desirable by every possible means to promote a kind and friendly intercourse between persons in different stations'. When the plan was finally made clear to him, Horner still declined involvement but suggested that Combe acquire 'the names of several liberal Tories as Councillors'.[52] This advice was relayed to

William Murray; one or two reputable reactionaries would 'serve well to shield us from Tory hostility'; 'Mr Horner said that he had found the greatest advantage in the School of Arts from having the countenance of a few of these gentlemen.'[53] Over the New Year holidays Combe worked assiduously to secure an apolitical public appearance for this most political of scientific enterprises. He invited Sir Henry Jardine, the Tory antiquarian and Councillor of the Royal Society of Edinburgh to participate, noting that Robison was a leading mover in the new scheme.[54] At the same time, he wrote to the Whig Lord Advocate, stressing the suitability of Nichol's political views: 'In his politics he is a Whig, bordering more on radicalism than Toryism . . . . [But he] conceives science to be the best antidote to fanaticism.'[55]

Combe's political tactics were clear. The Society for Diffusion was conceived as an instrument to further the social and cultural interests of the Edinburgh phrenologists. In order to advance those ends Combe had to secure for the Society a politically disinterested appearance. This he did by enlisting a number of men of different political allegiances, while ensuring that he and his allies retained effective control; and he solicited the participation of the "token" Tories and non-phrenologists by telling them what they wanted to hear about the new Society. This part of Combe's plan worked well. The first formal meeting of the Edinburgh Society for Aiding the General Diffusion of Science had been held on 21 December 1835 at Combe's house in Charlotte Square. The list of officers published in the New Year is evidence of Combe's astute politicking: the figurehead President was William Murray, the brother of the Lord Advocate; the two Vice-Presidents were Combe himself and Robison; the Council consisted of Sir George Stewart Mackenzie, the Tory *literatus* Sir Henry Jardine, and the naturalist-printer Patrick Neill; the Secretary-Treasurer was Combe's nephew the lawyer Robert Cox; and the Acting Committee in charge of lecturers comprised Nichol, the natural philosophy lecturer George Lees, the chemistry lecturers Dr D. B. Reid and Dr William Gregory, and Combe's brother, the physician Andrew Combe. Thus, a "disinterested" veneer was created, while all involved except Murray, Robison, Jardine and Lees were phrenologists or deeply sympathetic to the system.[56] Thus, the party-political manoeuvrings requisite to forestalling Edinburgh Tory and Whig opposition were successfully concluded.

Combe also obtained the support of a large group of Edinburgh scientific lecturers. Their reasons for participation in the new Society depended upon the structure of the local scientific career. For most of them, the pinnacle of a career in Scottish science and the end to which they constantly strove was a university chair, tenured *ad vitam aut culpam*. The great majority of Edinburgh men of science failed in that object, and thus lived in a continuing scramble to find suitable remunerative situations. It is only natural that we know far more about the few who succeeded than about the many who failed, yet the latter provide a more accurate portrait of the typical career patterns in a life of science.

Take, for example, the chemist David Boswell Reid (1805–63), a lecturer closely associated with Combe in the Society for Diffusion.[57] Reid graduated MD at the University of Edinburgh in 1830; but as early as 1828 he was acting as assistant to T. C. Hope, the Edinburgh Professor of chemistry, ultimately failing in a polemical campaign to have an independent chair of practical chemistry established.[58] During the mid to late 1830s alone, Reid supported himself by offering lecture and practical courses in chemistry and physiology to no fewer than twelve organizations in Scotland, many of them simultaneously: the Edinburgh School of Arts; the Edinburgh Philosophical Association; the Leith Mechanics' Institute; the Scottish Institution for the Education of Young Ladies; the Edinburgh High School, Academy and Lancasterian School; Heriot's Hospital, Edinburgh; the Argyle Square Medical School of Edinburgh; and the Scientific Associations of Arbroath, Dalkeith, Haddington, and Montrose; as well as his own extramural classes.[59] Until 1836 Reid had been pre-empted from positions in the Scottish Naval and Military Academy and the EPA by Andrew Fyfe and John Murray. Finally, he took Dr Johnson's 'noble road' south to London where he continued his lecturing and consulting career until he left for America in 1856 to become Medical Inspector to the United States Sanitary Commission. Similar struggles for scientific existence were fought by other lecturers on the new Society's Acting Committee: Dr William Gregory (until he became Professor of chemistry at King's College, Aberdeen in 1839, and five years later at Edinburgh), George Lees (who never secured a chair), and J. P. Nichol (developments in whose career in early 1836 had an important bearing on the fate of the Society for Diffusion). Nothing suited the lecturers' interests better than an organization which would publicize their talents and which put them in charge of filling paid provincial situations.

The revolt of the shopkeepers

The Edinburgh Society for Aiding the General Diffusion of Science was a brilliant idea. The party-political tactics required to bring it into being were deftly executed; the lecturers' energies were effectively harnessed. Yet, within months of its conception, the enterprise failed. There were several reasons for its failure, some involving particular personalities and the immediate vicissitudes of their careers, but the most significant cause of its demise was the hostile reaction of the Edinburgh Philosophical Association on whose reputation the new Society expected to capitalize.

Throughout 1835 relations between the Directors of the EPA and their "genius" George Combe had been deteriorating. On the one hand, the great man felt that the EPA's shopkeepers would not pay him his due, and, on the other, the shopkeepers took umbrage at what Combe considered to be his due. In the 1835 cheap 'people's edition' of *The Constitution of Man*, endowed by the Henderson Bequest, Combe had claimed credit for founding and inspiring the EPA.[60] This public claim to be their patron so offended the fiercely independent shopkeepers that they met in August to assert their view that the EPA had risen 'unpatronised, and hitherto un-encouraged by any notice from the great, or assistance from the learned'.[61] They refused to acknowledge Combe any longer as their presiding spirit. Neither did they recognize his financial assistance nor his priority in giving physiological lectures to a mixed-sex audience.[62] Moreover, the EPA was in dispute with Combe over payment for previous lectures.[63] His vanity insulted, Combe refused to give the introductory lecture in the coming session, throwing 'the Directors into disquietude', and forcing the reluctant Nichol to perform in his place.[64] Nichol was shocked by the merchants' insolence and presumption: 'A general vulgarity of mind – to be cured not in an hour or a day, but in a *generation* – which still characterizes most of that class, will explain much of what has happened, & ignorance will explain more.'[65]

The matters in dispute were not merely considerations of personal vanity or fiscal pettiness. Rather, they concerned the social and political control of cultural enterprises. Were these to be controlled by those groups they were ostensibly intended to serve, or were they to be directed by a local intellectual elite? This was why the EPA announced its total opposition to Combe's plan of a Society for Diffusion. Placing an advertisement in *The Scotsman* on

20 January 1836, the Directors of the EPA announced that they were in no way to be identified with Combe's enterprise.[66] No Director, they said, had been consulted in the scheme, 'although two of their present, and one of their former Lecturers, are to be found among its members'. Combe's 'manifesto', the Directors continued, 'conveys an indirect reflection on the Association for remissness of duty'. The EPA had, in fact, been diligently involved in the encouragement of similar provincial institutions and in the diffusion of science to the Edinburgh hinterlands.[67]

[The Directors] therefore cannot permit individuals to present themselves to the public as their patrons and encouragers, who, with very few exceptions, have refused to become members of the Association, and thereby to promote its objects.

In a telling phrase, the EPA referred to Combe and his new colleagues as 'this self-elected *Senatus Academicus*', thus tarring them with the same brush used against the resented intellectual elite of the University and the Royal Society of Edinburgh.

Within the EPA Combe's constant petty-bourgeois ally, the printer William Fraser, sprang to his defence. The new Society for Aiding the General Diffusion of Science was, Fraser said, a laudable body, aiming only 'to serve as a central medium of communication between town and country'; 'it was never dreamt that it could in the least interfere with the avowed object of the Edinburgh Philosophical Association – that of affording instruction to the inhabitants of Edinburgh'.[68] What then was the basis for the 'torrent of abuse' poured forth on the Society? It was, Fraser claimed, self-delusion about the scientific capabilities of the mercantile classes which made up the EPA's directorship. Local shopkeepers were indeed well-placed to assess local merchants' cultural wants, but men of 'high scientific and literary attainments', such as those who constituted the new Society, were required to plan the 'direction of the provincial scientific education of Scotland'.[69]

Combe and his associates, who had taken such pains to guard against attack by local political and cultural elites, had not evidently expected a *trahison des clercs et des épicières*. A phrenological gloss had to be put on their outburst:

[The EPA's] office bearers are chiefly shopkeepers; men highly respectable, but who by their office acknowledge themselves to stand *in need* of instruction and not to be qualified to communicate it. . . . Their advertisement is regarded by the members of our society as such a ridiculous manifestation of Self-esteem [phrenological organ no. 10] that we mean to take no notice of it.[70]

But to his eminent colleague Robison, Combe confessed his dismay about the Directors' reaction: 'What can have been the exciting cause to the extraordinary declaration of War issued against us by the Phil: Association? It surely has been the act of some intemperate individual who has altogether mistaken the nature of our Association.' He feared 'the harm they are likely to do to the advancement of Knowledge by continuing such manifestations'.[71]

Combe and Fraser bent their efforts towards securing a retraction.[72] Yet the Directors of the EPA remained obdurate. In April 1836 an acrimonious meeting of the EPA, instigated by Combe's allies while he was lecturing in Glasgow, failed to obtain a more accommodating response to the Society for Diffusion. Indeed, Combe's engagement as moral philosophy lecturer was terminated.[73] One unfriendly account even claimed that 'part of the audience', provoked by Combe's 'offensive' views on Christian doctrine and conduct in the Society for Diffusion, had 'hissed' his lectures that winter.[74] The EPA produced a pamphlet spelling out the grounds for its displeasure. First, it was upset that the new Society had attempted to trade upon the EPA's good name without prior consultation; this was felt to be unconscionable business practice.[75] More importantly, the Directors said that Combe's Society smacked of the very patronizing and monopolistic cultural politics that the EPA was dedicated to subverting: '[T]hey invite all provincial associations to consult them in the choice of lecturers, and all young men to court their patronage; a patronage which, were the Society successful, would soon be very extensive.'[76]

Worse, where the EPA had kept its men of science on a tight rein, the new Society proposed to put them back in control of their 'proper' masters. The consequences of this local Thermidorean reaction were clear: the Society 'would be above the public altogether'. Most pernicious of all was the plan to vest the accrediting of lecturers in an 'Acting Committee of Consultation and Examination' which consisted

of persons who were lecturers by profession, [and] whose interest was, therefore, opposed altogether to their duty, and who, from their very situation, were apt to crush the pretensions of all merit likely to rival their own, and to recommend only themselves, and such other persons as would submit to their usurped authority in the free republic of science.[77]

Finally, the Directors objected to the invidious social slur cast on the EPA by the composition of Combe's Society. They accused Fraser, in his bedazzlement by the great names of the Society for Diffusion, of betraying the EPA's principle of keeping the control of scientific culture in the hands of those whom it should serve:

In his admiration of the Baronets, Knights, Esquires, LL.D's, F.R.S.E.'s, &c. &c. &c., and his contempt for the "Messrs", Mr. Fraser had forgotten that his Society had, in their own Address, declared the "superior success" [of the merchants of the EPA as] the best judges of the qualifications of lecturers.[78]

By February Combe already feared that the EPA's actions had 'paralysed' his Society's 'efforts for a season'.[79] And, after the April debacle, he knew that the Society for Aiding the General Diffusion of Science was dead, and that his role in the EPA was at an end.[80] Later in 1836 *The Phrenological Journal* approvingly noted the foundation of the London Popular Educational Association whose Secretary was G. H. Lewes. 'This association,' it recorded,

appears to be formed on the model of the "Society for aiding the general Diffusion of Science", established in Edinburgh on 21st December 1835, but which was attacked so fiercely by the directors of the Philosophical Association, that its members dropt their scheme, and Scotland has, in consequence, been left without any central body for facilitating the arrangements between country audiences and qualified lecturers on science. We are happy to see that the plan has been adopted in London, and hope that no senseless ebullition of jealousy, from any quarter, will obstruct its usefulness in the sister kingdom.[81]

There is little doubt that opposition from the merchants of the EPA, animated by concern over the social control of scientific culture, was the major cause of the Society for Diffusion's demise. However, contingent circumstances affecting key individuals also contributed to its fate. In that week of April 1836 when the EPA was formulating its final "veto" of the Society, George Combe decided to apply for the vacant Edinburgh University chair of logic.[82] For many weeks until he was soundly thrashed in mid July by his "auld enemy" Sir William Hamilton in the Town Council election, Combe was immersed in the politicking and the soliciting of testimonials required by such a canvass. His capacity to sustain an effective war with the clerks and shopkeepers was drained. Moreover, from early in the year his colleague Nichol, the lecturer who had originally suggested the scheme and one of the Society's most energetic campaigners, was similarly involved in a contest for the Glasgow University Regius chair of practical astronomy.[83] Unlike Combe, Nichol was successful, receiving his Crown appointment on 6 February 1836. Immediately, he threw himself into preparations for his duties, renovating the Glasgow Observatory and devising schemes for more effective academic and popular instruction in astronomy. Thus Nichol, too, was lost to the Edinburgh cause.[84] There was no one left with both the commitment and the capability to persist with the Society for Diffusion in the face of local opposition. Thus, an idea for an organization which seemed "right for the times", the strategic politics of which had been in many respects so well worked out, and which promised to unite the social interests of diverse local groups, utterly failed.[85]

Class and scientific culture in a provincial metropolis

In the current state of the social history of British science a highly detailed empirical focus is often found associated with antiquarianism, and, conversely, pretensions to theoretical profundity with empirical vacuity. However, certain generally valuable historiographic perspectives can only be sustained by the use of a high-resolution microscope. This study has been concerned to show the diffusion of science as a political process. Yet the range of relevant political factors would not have been visible had the empirical focus been somewhat fuzzier. Had one scrutinized these materials less closely, had the fine structure of political transactions not been displayed, the diffusion of science might well have appeared to proceed by its own immanent "logic". "Lazy" historical practice helps to underwrite "passive" views of cultural transmission. In drawing the appropriate conclusions, let us therefore proceed from the particular to the general.

The particular local context we have examined was a *provincial metropolis*. In one very obvious sense this meant that Edinburgh produced culture for itself, for its export market, and for the consumption of its hinterlands. Intellectuals socialized into the metropolitan values of their training, pressed in upon metropolitan institutions and created an intensely competitive "buyers' market" for cultural display, some of the consequences of which we have seen in the EPA and the Society for Diffusion.

More generally interesting perspectives can be developed by asking how a provincial–metropolitan context bore upon cultural behaviour in the class structure. One immediately notices that different social classes maintained different views about the nature, social uses, and organization of scientific culture.[86] Intellectual elites saw in science an acutal or potential career or an appropriate avocation. Other social groups viewed science as practically useful knowledge, a legitimation of the existing social order, or an important resource for undermining that order. Divergent social interests gave rise to conflict in the wider society and corresponding divergent interests in culture gave rise to conflict in science and its institutions.

If one looks at scientific culture in its early nineteenth-century British context, it is tempting to see national divides between groups with different social interests in science, for example between the purposes of the intellectual elites of the British Association for the Advancement of Science and those meliorist materialists of the Combe circle. However, the study of the diffusion of scientific culture in and from a provincial metropolis provides a quite different perspective. What one sees in Edinburgh is conflict between groups which had a *national* (or international) cultural horizon and those whose cultural interests were almost purely *local*. A provincial metropolitan context harboured both sorts of group and sharpened tensions between them which existed elsewhere in attenuated form. Moreover, the kinds of conflicts and alliances between groups, which the provincial metropolis fostered, illustrate some interesting points about the notion of class appropriate to the social history of science generally.

The Combeites, for example, attempted, and to an extent succeeded, to form social and cultural alliances with the lesser bourgeoisie, and to mobilize them around reformist social goals. They did this in opposition to the recognized intellectual elites of Edinburgh. Yet, as we have seen, they fell out with the merchants and shopkeepers who were their "natural" allies and formed tactical alliances with elite individuals, such as John Robison, who were their "natural" enemies. The major cause of these unexpected formations was that the cultural ambitions of *both* Combeites *and* persons like Robison were national, while those of the EPA's petty-bourgeoisie were local. Given the specific environment of a provincial metropolis and the particularity of tactical aims, the arrangement of social groups in cultural enterprises followed patterns which could not be predicted from a view of classes which conceived them as "strata" with fixed social interests and "logically entailed" cultural orientations.

What arises, therefore, from an examination of cultural behaviour in a provincial metropolis is a valuable historiographic orientation towards a conception of class. Classes are *not* like geological strata, layered in society, whose relations with each other are rigidly fixed. Similarly, the cultural behaviour of classes is not found to be locked into patterns theoretically "appropriate" to their position in the "strata". The more closely one looks at the politics of culture in a concrete context, the more one observes the finely textured dynamic processes by which men create their own social environment and tell the historian what class is. To write better social history of science we have to write better, and more sensitive, social history. Thus, it is one of the most acute social historians who has provided us with the most apposite notion of class; class, writes E. P. Thompson, is not 'a "structure", nor even a "category", but. . . something which in fact happens. . .in human relationships. . . . Class is defined by men as they live their own history'.[87]

Acknowledgements

For comments on earlier drafts of this paper, I am grateful to Paul Baxter, David Miller, and Roy Porter. I thank the staffs of the National Library of Scotland (NLS), the Edinburgh University Library (EUL), and the Edinburgh Room of the Edinburgh Public Library (EPL) for their help. In the Notes and references I have indicated by the appropriate abbreviation the location of rare published items. For permission to quote from manuscripts in their care, I am grateful to the Trustees of the NLS.

Notes and references

1   An excellent study of scientific diffusion as an historically situated and active process is M. C. Jacob, *The Newtonians and the English Revolution 1689–1720*, Hassocks, Sussex, 1976, chs. iv–v; for historiographic treatment, S. Shapin, 'Social uses of science', in R. S.

Porter and G. S. Rousseau (eds.), *The Ferment of Knowledge: Studies in the Historiography of Eighteenth-Century Science*, Cambridge, 1980, 93–139.

2   For science in the University of Edinburgh, J. B. Morrell, 'The University of Edinburgh in the late eighteenth century: its scientific eminence and academic structure', *Isis*, 1971, **62**, 158–71; *idem*, 'Science and Scottish university reform: Edinburgh in 1826', *British Journal for the History of Science*, 1972, **6**, 39–56; *idem*, 'The patronage of mid-Victorian science in the University of Edinburgh', *Science Studies*, 1973, **3**, 353–88.

3   A. Ford (ed.), *The 1826 Journal of John James Audubon*, Norman, Oklahoma, 1967, p. 256.

4   Tom Stoppard, *Jumpers*, London, 1972, although the slur is omitted from the published version of the play. For the Edinburgh "declinist" lament, H. Cockburn, *Life of Lord Jeffrey*, 2 vols., Edinburgh, 1852, **i**, pp. 157–61; [R. Mudie], *The Modern Athens*, 2nd edn, Edinburgh, 1825, pp. 210, 221–3; and J. R. R. Christie, 'The rise and fall of Scottish science', in M. Crosland (ed.), *The Emergence of Science in Western Europe*, London, 1975, pp. 111–26.

5   H. Cockburn, *Memorials of his Time*, new edn, Edinburgh, 1910, pp. 164–5.

6   H. Cockburn, *Journal of Henry Cockburn...1831–1854*, 2 vols., Edinburgh, 1874, **i**, p. 117.

7   There is no modern study of the Edinburgh School of Arts, but see S. Shapin and B. Barnes, 'Science, nature and control: interpreting mechanics' institutes', *Social Studies of Science*, 1977, **7**, 31–74.

8   The best contemporary account is H. Cockburn, op. cit. (6), **i**, pp. 1–43. For an excellent modern treatment, L. J. Saunders, *Scottish Democracy 1815–1840: The Social and Intellectual Background*, Edinburgh, 1950.

9   Cockburn, op. cit. (6), **i**, pp. 73–4. By 1835 the Association's name, felt to be 'as long as a Dutchman's', was changed to the 'Edinburgh Philosophical Association', and this name is used henceforth. There is no modern account of the EPA. The key references for its founding and early history are cited in notes, 11, 12, 14, 19, 21, 36, 60, 62, 67, and 75 below; for its later history, note 85.

10   For the British petty-bourgeoisie see G. Crossick (ed.), *The Lower Middle Class in Britain 1870–1914*, London, 1977, esp. ch. i; *idem, An Artisan Elite in Victorian Society: Kentish London, 1840–1880*, London, 1978, esp. pp. 130–8 for relations between labour aristocracy and lower middle class; R. Q. Gray, *The Labour Aristocracy in*

Victorian Edinburgh, Oxford, 1976; G. Anderson, *Victorian Clerks*, Manchester, 1976.

11 James Simpson, *The Philosophy of Education*, 2nd edn, Edinburgh, 1836, p. 180. Appendix IV to this work (pp. 238–49) contains a 'Summary of the proceedings of the Edinburgh Philosophical Association, from its Institution in 1832 to June 1836'.

12 The list of EPA Directors for 1832-3 is reproduced in ibid., p. 180, and in *[Second] Report by the Directors of the Edinburgh Association for Useful and Entertaining Science*, Edinburgh, 1833, p. 6 [NLS].

13 Simpson, op. cit. (11), p. 244.

14 *Popular Education. Address to the Public, by the Directors of the Edinburgh Philosophical Association*, Edinburgh, 1835, p. 8 [NLS, EUL].

15 'Edinburgh Philosophical Association', *The Scotsman*, 10 October 1835.

16 *Popular Education*, op. cit. (14), p. 6.

17 'Popular lectures in chemistry', *The Scotsman*, 12 January 1833.

18 *Popular Education*, op. cit. (14), pp. 5–6.

19 *[First] Report by the Committee for Arranging the Preliminary Details of the Courses of Lectures on Geology, Chemistry, and Phrenology-...during the Winter 1832-3*, Edinburgh, 1832, pp. 3–4 [NLS]; *Popular Education*, op. cit. (14), p. 9.

20 *[Second] Report*, op. cit. (12) p. 3.

21 *Third Report by the Directors of the Edinburgh Association for Procuring Instruction in Useful and Entertaining Science*, Edinburgh, 1833, p. 2. [NLS].

22 Simpson, op. cit. (11), p. 245; cf. *Popular Education*, op. cit. (14), p. 10.

23 For Combe's Edinburgh circle, D. A. De Giustino, 'Phrenology in Britain, 1815–1855: a study of George Combe and his circle', Ph.D thesis, University of Wisconsin, 1969, which is better on Edinburgh than *idem, Conquest of Mind: Phrenology and Victorian Social Thought*, London, 1975. For Edinburgh phrenology, G. N. Cantor, 'The Edinburgh phrenology debate: 1803–1828', *Annals of Science*, 1975, **32**, 195–218; S. Shapin, 'Phrenological knowledge and the social structure of early nineteenth-century Edinburgh', ibid., 219–43; *idem*, 'The politics of observation: cerebral anatomy and social interests in the Edinburgh phrenology disputes', in R. Wallis (ed.), *On the Margins of Science: The Social Construction of Rejected Knowledge, Sociological Review Monographs*, 1979, **27**, 139–78; *idem*, 'Homo phrenologicus: anthropological perspectives on an

historical problem', in B. Barnes and S. Shapin (eds.), *Natural Order: Historical Studies of Scientific Culture*, Beverly Hills and London, 1979, pp. 41–71.

24   C. Gibbon, *The Life of George Combe*, 2 vols., London, 1878, i, p. 177.

25   ibid., i, pp. 235–6.

26   ibid., i, p. 254; *[First] Report*, op. cit. (19), p. 1; *Popular Education*, op. cit. (14), p. 1; *Additional Testimonials on Behalf of George Combe, as a Candidate for the Chair of Logic in the University of Edinburgh*, 1836, p. 121 [EUL]; W. F[raser], *Hints on the Unlimited Diffusion of Useful Knowledge, at No Expense to the Reader, through the Medium of the Mercantile and Trading Classes*, Edinburgh, 1834 [NLS].

27   For phrenology in British popular science, R. Cooter, 'The cultural meaning of popular science: phrenology and the organization of consent in nineteenth-century Britain', Ph.D thesis, University of Cambridge, 1978.

28   Gibbon, op. cit. (24), i, p. 254.

29   For this Society, Simpson, op. cit. (11), pp. 256–63.

30   W. B. Hodgson, *Lecture on Education, Delivered in the Freemasons' Hall, at the Opening of the Second Session of the Edinburgh Association of the Working Classes, for their Social, Intellectual, and Moral Improvement, Monday 16th October 1837*, Edinburgh, 1837, pp. 33–5, 46–8 [NLS].

31   Gibbon, op. cit. (24), i, pp. 254–64.

32   'Miscellaneous notices', *The Phrenological Journal and Miscellany*, 1836–7, **10**, 383. Hodgson became the first Edinburgh Professor of political economy in 1871.

33   *Testimonials on Behalf of George Combe, as a Candidate for the Chair of Logic in the University of Edinburgh*, Edinburgh, 1836 [EUL]; *Additional Testimonials*, op. cit. (26); Gibbon, op. cit. (24), i, pp. 318–31.

34   Gibbon, op. cit. (24), ii, pp. 101, 115, 134, 360; 'The Phrenological Association', *The Phrenological Journal*, 1839, **12**, 29–35, 201–6, 403–11; G. S. Mackenzie, 'The split in the Phrenological Association', ibid., 1842, n.s. **5**, 343–6.

35   For example, 'Combe's Constitution of Man', *Presbyterian Review*, 1836–7, **9**, 92–118, 429–50 [EUL].

36   "Philomathes" [W. Pyper], 'Phrenological system of education', *Edinburgh Advertiser*, 24 December 1833 and 31 December 1833 [EPL]; and the EPA's reply to these charges in *Fourth Report by the*

Directors of the Edinburgh Association for Providing Instruction in Useful and Entertaining Sciences, Edinburgh, 1834, pp. 2–5 [NLS]. On the Kirk's reaction: P. Baxter, 'Combe, *The Constitution of Man*, and the churches', paper read to British Society for the History of Science, July 1978, Durham.

37   Cockburn, op. cit. (6), **i**, p. 63.

38   ibid., **i**, p. 64.

39   Robison to Combe, 3 September 1834, NLS 7233, fo. 65; Robison to Combe, 20 March 1835, NLS 7236, fo. 63.

40   Fraser to Combe, 9 January 1835, NLS 7235, ff. 25–6; Nichol to Combe, 4 January 1835, NLS 7236, ff. 3–4.

41   'Mr. J. P. Nichol', *The Scotsman*, 24 October 1835; R. Cox to Combe, 27 October 1835, NLS 7234, ff. 175–6.

42   Nichol to Combe, 9 October 1835, NLS 7236, ff. 5–6.

43   Combe to Mackenzie, 6 December 1835, NLS 7386, ff. 432–3.

44   Nichol to Combe, 9 October 1835, NLS 7236, ff. 5–6.

45   Nichol to Combe, 4 January 1835, NLS 7236, ff. 3–4.

46   On Robison, J. D. Forbes, 'Biographical notice of the late Sir John Robison, K. H., Sec. R. S. Ed.', *Proceedings of the Royal Society of Edinburgh*, 1851, **2**, 68–78.

47   'Society for aiding the General Diffusion of Science', *The Scotsman*, 9 January 1836; 'Address by the Edinburgh Society for Aiding in the General Diffusion of Science', *The Phrenological Journal*, 1836–7, **10**, 36–9. Tens of thousands of prospectuses were supposedly distributed throughout the provinces, but no copies apparently survive. See also note 67 below.

48   Combe to Mackenzie, 6 December 1835, NLS 7386, ff. 432–3.

49   Combe to W. Murray, 14 December 1835, NLS 7386, ff. 439–41.

50   Combe to Mackenzie, 17 December 1835, NLS 7386, ff. 443–4.

51   Horner to Combe, 31 December 1835, NLS 7235, ff. 91–2.

52   Combe to Horner, 31 December 1835, NLS 7386, fo. 444.

53   Combe to W. Murray, 30 December 1835, NLS 7386, fo. 455.

54   Combe to Jardine, 1 January 1836, NLS 7386, fo. 457.

55   Combe to J. A. Murray, 11 January 1836, NLS 7386, ff. 466–7.

56   The list of officers is printed in the two sources in note 47.

57   H. Reid, *Memoir of the Late David Boswell Reid*, Edinburgh, 1863; M. F. Conolly (ed.), *Biographical Dictionary of Eminent Men of Fife*, Cupar, Fife, 1866, p. 377.

58   J. B. Morrell, 'Practical chemistry in the University of Edinburgh 1799–1843', *Ambix*, 1969, **16**, 66–80; *Testimonials Regarding Dr. D. B. Reid's Qualifications as a Lecturer on Chemistry, and as a Teacher*

of Practical Chemistry, Edinburgh, 1833 [EUL]. Reid's opponent on that occasion was Dr Andrew Fyfe, the other major extra-mural chemistry lecturer in Edinburgh and Reid's constant competitor for vacant positions.

59   The list is culled from many newspaper notices, and especially D. B. Reid, *Rudiments of Chemistry*, Edinburgh, 1836, p. vi, for his current engagements.

60   George Combe, *The Constitution of Man*, Henderson edn, Edinburgh, 1835, Appendix v; cf. 4th edn, Edinburgh, 1836, Appendix x.

61   *Popular Education*, op. cit. (14), p. 8.

62   W. Fraser, *Reasons of Protest against the Laws of the Edinburgh Philosophical Association*, Edinburgh, 1836, p. 12 [NLS]; Cox to Combe, 20 October 1835, NLS 7234, ff. 173–4; J. P. Falkner to Combe, 15 October 1835, NLS 7234, ff. 17–18; Falkner to Combe, 26 October 1835, NLS 7234, ff. 19–20; Nichol to Combe, 17 October 1835, NLS 7236, ff. 7–8.

63   Combe to J. Dun, 14 September 1835, NLS 7386, ff. 382–3; Combe to Falkner, 15 September 1835, NLS 7386, ff. 388–9.

64   Nichol to Combe, 9 October 1835, NLS 7236, ff. 5–6.

65   Nichol to Combe, 17 October 1835, NLS 7236, ff. 7–8.

66   'Edinburgh Philosophical Association', *The Scotsman*, 20 January 1836; advertisement reproduced in Fraser, op. cit. (62), pp. 19–20.

67   For evidence that the EPA provided the model for similar organizations in the Scottish provinces, see testimony of James Simpson in *Report from the Select Committee on Education in England and Wales*, Parliamentary Papers, 1835, **7**, 202.

68   Fraser, op. cit. (62), p. 10; Combe to W. Murray, 23 March 1836, NLS 7386, ff. 526–7, where Combe denied that he had anything to do with the composition of Fraser's pamphlet.

69   Fraser, op. cit. (62), p. 11.

70   Combe to D. G. Hallyburton, 30 January 1836, NLS 7386, ff. 476–8.

71   Combe to Robison, 21 January 1836, NLS 7241, fo. 42.

72   Among the many pertinent letters, Nichol to Combe, 19 January 1836, NLS 7240, ff. 190–1; M. Lothian to Combe, 28 April 1836, NLS 7239, ff. 141–2; Combe to Lothian, 1 June 1836, NLS 7387, fo. 17; Fraser to Combe, 5 April 1836, NLS 7238, ff. 193–4; Cox to Combe, 4 April 1836, NLS 7238, ff. 79–80.

73   These lectures were ultimately published as G. Combe, *Moral Philosophy; or the Duties of Man Considered in his Individual, Social and Domestic Capacities*, Edinburgh, 1840.

74   'Edinburgh', *Scottish Guardian* [Glasgow], 12 April 1836; printed,

with Combe's reply, in G. Combe, *The Suppressed Documents: or, an Appeal to the Public against the Conductors of the 'Scottish Guardian'*, Glasgow, 1836, pp. 4–7 [NLS]. I owe these references to Paul Baxter.

75 *Answer of the Directors of the Edinburgh Philosophical Association to the Protest of Mr. William Fraser, Printer* . . . , Edinburgh, 1836, p. 11 [NLS].

76 ibid., p. 12.

77 ibid.

78 ibid., pp. 14–15.

79 Combe to W. Tait, 16 February 1836, NLS 7386, ff. 495–6.

80 Combe to A. J. D. Dorsey, 25 May 1836, NLS 7387, ff. 11–12.

81 *The Phrenological Journal*, 1836–7, **10**, 499–500.

82 Gibbon, op. cit. (24), **i**, pp. 318–31.

83 Combe to J. A. Murray, 11 January 1836, NLS 7386, ff. 466–7; Combe to Nichol, 11 February 1836, NLS 7386, fo. 490.

84 James Coutts, *A History of the University of Glasgow* . . . , Glasgow, 1909, pp. 388–9. For Nichol, P. Baxter, 'Science and belief in Scotland, 1800–1860', Ph.D thesis, University of Edinburgh, forthcoming.

85 A brief ironic note is in order about the later fate of the EPA. The petty-bourgeoisie was quite right to be suspicious about the intentions of local elites. One or two lawyers who had joined before the Combe affair quickly exerted their influence on the social organization and cultural proceedings of the EPA. The EPA first became less scientific, then more socially elevated, until finally in 1846 it was reconstituted as 'The Philosophical Institution', a kind of Athenaeum dominated by lawyers and professors and thoroughly respectable. The minute books of the Philosophical Institution are in the Edinburgh Room of the Edinburgh Public Library, whose catalogue includes an entry for minutes from the year 1836 which have never been located and must be presumed lost or mistakenly entered. For the transformation of the EPA into the Philosophical Institution, *Prospectus of the Philosophical Institution to be Established in Edinburgh*, Edinburgh, 1846 [EPL, EUL]; *Syllabus of Lectures to be Delivered before the Edinburgh Philosophical Association* . . . *Session 1845–46*, Edinburgh, 1845 [EUL]; *The Jubilee Book of the Philosophical Institution*, Edinburgh, 1897; W. A. Miller, *The 'Philosophical': A Short History of the Edinburgh Philosophical Institution* . . . *1846–1948*, Edinburgh, 1949.

86 For studies building on this perception: A. Thackray, 'Natural

knowledge in cultural context: the Manchester model', *American Historical Review*, 1974, **79**, 672–709; S. Shapin, 'The Pottery Philosophical Society, 1819–1835: an examination of the cultural uses of provincial science', *Science Studies*, 1972, **2**, 311–36; I. Inkster, 'Science and society in the metropolis: a preliminary examination of the social and economic context of the Askesian Society of London, 1796–1807', *Annals of Science*, 1977, **34**, 1–32.

87   E. P. Thompson, *The Making of the English Working Class*, Harmondsworth, 1968, pp. 9, 11.

6 Science in a commercial city: Bristol 1820–60

Michael Neve

The investigation of provincial and metropolitan scientific cultures in the nineteenth century is at the same time an investigation into the varieties of local social history. Physical geography and cultural geography were inextricably involved with each other, and the two dimensions were bound together and explicated by the idiosyncracies of local social structure. The movement down through the strata of historical generalizations has already gone far.[1] There remain however certain possibilities that require study.

Science and its cultural uses can still, for perfectly proper historical reasons, be linked in the early nineteenth century to two main historical vehicles: Dissent and Utility. In the case of the relation between science and dissenting movements, it has been suggested that a fertile scientific culture was generated among marginal men as a means of social legitimation. In the case of Utility, science has been shown to have been of use to a variety of distinct social groups, most famously to improving landowners in the lowlands of Scotland for instance, and through the institution of metropolitan societies which were made to serve, however briefly, the purposes of landed classes.[2] In his study of the Royal Institution, Berman has broken open the question of provincial and metropolitan relations by showing how an urban elite only gradually came to control an institution actually placed in the heart of the metropolis, and has related this to the gradual strengthening of urban, utilitarian, professional power in nineteenth-century Britain in general.

It is possible to add further to these striking historical insights. A dimension that has not received sufficient examination, despite researches into 'social control' and other ideological issues, is that of scientific *conservatism*, where science plays a decisive role *within* established elites in the manufacture of a new language of authority and political power.[3] Here, the position and purposes of provincial, socially established, bourgeoisies and their relation to promoting an effective social order are of interest. Science, in some provincial contexts, was not produced through radicalism and marginalism, nor was it a culture form that made a passage from "outsider" to "insider" between say 1780 and 1840. Instead, in centres such as Bath and Bristol, it developed inside already powerful and confident social elites, whose *political* task in the period 1780–1830 was a prolonged attempt to solidify power with the still hegemonic ruling landed class. "Bourgeois science" can then be investigated as part of that nervous historical collaboration between bourgeoisie and aristocracy that has engaged so many political historians of the nineteenth century.[4]

The theme of this essay then is that far from posing threats to the standards of established culture and religion, "science" was a system that developed *within* that culture with the purpose of strengthening it. This would certainly be the conclusion of a historical perspective taken from Trinity College Cambridge rather than from Manchester or Newcastle. As Thackray has argued, it was also the case that science acted as a stepping stone for dissenting families in Manchester to achieve the status of Anglican landowner. But the process of assimilation or penetration (the choice of word is a political one) occurred not only earlier but also at different speeds in older centres and in the newer industrial provincial centres. The adoption of natural philosophy was not always a move from marginality to centrality, or radicalism to conservatism. Instead, it formed, in the case of Bristol and elsewhere, one part of the efforts of an already well-established social elite to maintain its domination, through the development of scientific languages and institutions. It can even be argued that certain political differences, such as Whig and Tory, became acknowledged as outmoded: if maintained, they were used to give the *appearance* of political conflict. The more urgent task was the coining of a language of natural power that was also a political language.[5] In certain examples, of which Bristol is one, the casting aside of relatively uncontroversial oppositions applied to elite religious affiliations as well.

Two fruitful areas of research for developing these ideas may be suggested. First, a detailed study of the number of 'conversions' of dissenting individuals into established religion as a response to the destruction of the French *ancien régime*. How widespread was this? Second, what devices were at hand for promoting, through science, a political collaboration between bourgeoisie and aristocracy? One good candidate was the British Association for the Advancement of

Science. The BAAS may well have been the perfect, migratory common ground upon which provincial bourgeoisies met, to maintain an essential aristocratic connection, and of course to exclude all voices that failed to contribute to the crystallization of the new conservatism. Whether or not this *rapprochement* was successful, it is surely useful to recognize the highly conservative, but far from moribund, aspects of early nineteenth-century scientific culture. What is needed is not the sociology of exclusion, but the evidence, from science as well as politics, for the formidable conservative achievement of Sir Robert Peel.

In terms of the liveliness and duration of its scientific culture, the mercantile centre of Bristol offers a much slighter story than Edinburgh, Newcastle, Manchester and London.[6] Its wholehearted conservatism was of a type with say, Bath or York, but at odds with the developing industrial areas. Scientific activities in Bristol were never prolonged or particularly successful. Bristol was a commercial Atlantic port which from 1800 onwards was entering a period of relative decline. It produced a short-lived scientific institution, as well as certain other scientific activities, which remained viable until the middle of the century. From 1850 until about 1880, the main impetus for scientific and intellectual inquiry came from the city's medical men, although their domination was not so obvious in the earlier period. The latter part of the nineteenth century saw a new stage, as well as certain new forms, of local intellectual culture.[7]

The scientific culture of early nineteenth-century Bristol was markedly non-utilitarian, and conservative. It was not the product of marginal men, but rather the brief achievement of a well-established, predominantly Anglican, bourgeoisie, the sources of whose wealth had been secured for at least two generations. This Anglican domination was tinged with an active element of elite Unitarians, centred on the famous Lewin's Mead meeting. Some of the Unitarian families were more powerful and established than their Anglican merchant peers. The chief scientific institution established by this group, the Bristol Institution for the Advancement of Science (1823) was highly exclusive, set as it was in a city many of whose other elite worlds – most famously the Corporation – were notoriously aloof. That Corporation was dominated by Tories from 1812 onwards, although the two member seat of Bristol was effectively shared by the winning candidates, even when ostensibly politically opposed.

A notable aspect of the history of Bristol, and its neighbouring town of Bath, was the absence in the heyday of the eighteenth-century "Enlightenment" of organized scientific culture.[8] Instead, the cultural developments that did take place – assembly rooms, libraries, hospitals, charities, and the visits of itinerant lecturers – display two notable features. They were arranged around a tourist economy, subject to seasonal variation and dominated by visitors, not natives. Even more importantly, they were *imitative* of the metropolis. West Country culture in both Bristol (at its spa at the Hotwells and in the polite suburb of Clifton) and Bath mimicked the metropolis in the eighteenth century.

The development of scientific culture in Bristol in the nineteenth century bore the marks of this imitativeness. It would be wrong to take the obvious cultural differences between these two periods, most especially the marked evangelical atmosphere of the nineteenth century, as an indication of cultural independence. The imitative relation to the senior cultural modes of the realm remained wholly intact, even though ballroom gave way to Bible society. Both Bristol and Bath may well have been trapped in a dependence on metropolitan forms which only industrialization could break, and industrialization left few marks on the economy of the West Country. A distinct local culture for Bristol had to wait until at least the third quarter of the nineteenth century.[9] As far as the period 1820–60 is concerned, the Bristol social elite developed institutions and contacts with a reawakened evangelical movement and with the forms and commitments of an alert conservative establishment. Bristol's scientific culture was a secondary formation, based on the need of its proponents to attach themselves to the new conservatism of post-Napoleonic Britain.[10]

The economic foundations for the financing of science and scientific institutions in Bristol in 1820 came from eighteenth-century roots.[11] Bristol had been the centre of trade and communications in the West of England, and its main activities were: its port; the production of brass and iron, often for shipping; derivatives of the West Indian trade, especially sugar (there were at the end of the eighteenth century more than twenty refineries); and alongside these, the lesser activities of glass production, soap, porter, tobacco, lead shot, and chocolate. Banks abounded, often owned by families with accompanying interests in commercial enterprises. The city boasted a large number of charities and philanthropies; an Infirmary founded in 1736; an idiosyncratic

'Corporation of the Poor' with its combined lunatic asylum and poorhouse, St Peter's Hospital; a cathedral that had Joseph Butler as Bishop in the 1730s; and a residential suburb, Clifton, that became the byword for a salubrious, separate village environment, away from the city itself.

By 1800 the relative decline of Bristol's economy began to make itself felt. Population had increased to near 100,000 but the mercantile economy was weakening.[12] Attempts to improve harbour facilities to deal adequately with the wide tidal range of the Avon (later to prove of scientific interest) exposed weaknesses and conflicts between the Dock Company, the Society of Merchant Venturers and the notoriously self-protective Corporation of Bristol.[13] The taxes payable on ships and goods coming into Bristol were considerably higher than those levied in London and Liverpool. Bristol had become too dependent on West Indian and Irish trades, themselves beginning a steep economic decline, and a post-war mini boom could not conceal severe structural faults. In areas of manufacturing such as glass, the relative decline was emphasized by developments elsewhere in Britain, notably Lancashire. There was no industry in Bristol comparable to the new industries, such as cotton, in the North.

An elite of local merchants, whose wealth was based on sugar, brass, iron founding, lead shot and banking, and the West Indian trade itself, experienced difficulties in the period 1820–40. Families such as the Harfords (banking, brass), the Vaughans (merchants), the Daubenys (banking, sugar and general commerce), the Ricketts (glass), the Acramans (iron), the Fripps (sugar, candles, soap), the Miles (merchants) and the Georges (brewing) were all locked together in a "commercial freemasonry", and on the point of sharing a collective downward fate. Alford suggests that none the less

once an individual businessman has achieved a solid level of prosperity, profit maximisation usually becomes an 'inferior' good. Both glass and tobacco provide examples of later generations of established families withdrawing altogether from business during the 1820s and 1830s in order to follow more gentlemanly pursuits, and, from what is known of others of their kind, they appear quite typical.[14]

It was from this equivocal base, from sources of wealth that had peaked and now sought only their own self-protection, that most local scientific activity up to 1850 came. The decline of this clique, and the decline of the scientific ideology that it promoted, marks off the economic and cultural history of Bristol in the first half of the nineteenth century from that of the second half.

Before the establishment of the Bristol Institution in 1823, local scientific culture consisted of sporadic lectures and exhibitions chiefly by itinerants.[15] There also existed from about 1809 a Literary and Philosophical Society which met in St Augustine's Place and then at no. 1 Trinity Street. Lecturers here included the geological writer Robert Bakewell (1768–1843), then lecturing at the Russell and Surrey Institutions in London, John Jackson also of the Surrey Institution, Robert Addams of the Royal Institution, and one Bristol activist, Michael Fryer FSA, teacher of mathematics and secretary to the Society. Subscriptions to put the Society on a sound footing appear to have reached between £5000 and £6000 by 1809,[16] and fluctuated in the course of the society's quirky history. Sporadic lecturing in the city had also been provided by some local medical men, including the celebrated Thomas Beddoes (1760–1808) who had been assisted in lectures on anatomy and chemistry in 1797 and 1798 by two local surgeons, Richard Smith and Francis Cheyne Bowles. Beddoes was at the centre of an idiosyncratic group based around his Pneumatic Institute in Clifton, and no institution grew out of his various medical activities. Attempts to set up a permanent room 'where all branches of philosophy might be given' had been countermanded by the Dean and Chapter of Bristol Cathedral, partly because of Beddoes's political tendencies; in any event, Beddoes's Jacobin skills did not extend to making his anatomical and chemical explanations comprehensible to his audience.

Developing alongside this semi-organized scientific activity (activity that in Beddoes's case was based on a network taking in the Midlands, Birmingham and the Continent as much as at Bristol), there also began that rash of philanthropic concerns that were the mark of bourgeois nervousness resulting from the events in France. Bristol now added to its already large number of established charities the Bristol Auxiliary Bible Society (1810); the Bristol Church of England Missionary Society (1813); there was increased support for the Bristol Diocesan schools; the Church of England Tract Society, Bristol branch was founded in 1811; the Prudent Man's Friend Society in 1813, and the Bristol Union Society for the Promotion of Sunday Schools in 1814. Finally, the Bristol Savings Bank was started in August 1817. Throughout this period, the city's elite gathered at endless meetings calling for the defence of the Protestant constitution, there to toast its one-time MP Edmund Burke, to stress the need to increase charity in times of economic distress, and to subscribe heavily to the British and Foreign Bible Society. The sense of threat from the city's Methodist groups and from the disturbances in the mining communities of North Somerset and Kingswood was strong. A history has yet to be written of the powerful menace of the bandit gang who raided from Cockroad, near Kingswood, before being imprisoned and "educated". The Bridge Riots of 1793, the Reform Riots of 1831 and the pauper riots in St Peter's Hospital in 1832 – all speak of a violent popular politics. Not surprisingly, Bristol's share of Lord Liverpool's £1 million grant of 1818 for the building of Anglican churches was spent on a church put up in the alien proletarian world of St Philip's Marsh.

The final securing of funds to establish a scientific institution coincided with these events. The Bristol Institution for the Advancement of Science, Literature and the Arts was effective from 1823, the same year as the new Chamber of Commerce was opened. There is a clear overlap between those who financed the philanthropic, commercial and cultural activities that began in the period 1810–25. Apart from the families mentioned as part of the commercial "inner circle", the core financial group was dominated by Tory Anglicans, including D. W. Acraman, ironfounder and patron of the local artistic community; Thomas Daniel (1763–1854), senior civic figure and West Indian merchant; John Kerle Haberfield (1785–1857), attorney; Christopher George (1786–1866), lead shot manufacturer, and Philip and Richard Vaughan (1767–1833). The leading Whigs associated with the Institution were sociologically indistinguishable from this group, and included the West India proprietor, Richard Bright (1756–1840) whose third son, Richard Bright (1789–1858), was a Fellow of the Geological Society of London and discoverer of Bright's disease. Like Richard Bright senior, another local Whig, Michael Hinton Castle (1785–1845) was a Unitarian, and a distiller. In some leading Unitarian families, the Ricketts and the Lunells, there were conversions to Anglicanism. But in no sense was the Unitarianism founded on the Lewin's Mead Meeting an outsider culture. Bristol Unitarianism, from the time of John Prior Estlin (1747–1817) to well beyond the end of the ministry of Lant Carpenter (1780–1840) was an intrinsic part of upper-class activity within the city.

One important spur to the development of scientific culture was noted by the Tory newspaper editor John Mathew Gutch (1776–1861), of Christ's Hospital and friend of Coleridge and Lamb, in a letter in his own paper in October 1822. Gutch voiced the fear that commercial life and science were antithetical, but pointed out that a number of figures had 'risen up' in Bristol who could bring science to new and beneficial heights.[18] Gutch was thinking specifically of James Cowles Prichard, ethnologist and physician at the Bristol Infirmary; William Daniel Conybeare, then lecturer in the nearby parish of Brislington, and geologist; and George Cumberland (1754–1848), antiquarian, geologist, patron of William Blake and retired insurance broker. It was time, said Gutch, to finalize the plans for an institution to house these talents. Another important organizational figure then in the city was Henry Beeke (1751–1837), Dean of Bristol, product of the home of liberal Anglicanism, Oriel College Oxford, geologist and Regius Professor of modern history at Oxford, 1801–13. Beeke involved himself in scientific and clerical pursuits, and was also chairman of the Bristol Savings Bank from August 1817.[19] The presence of these men involved the Bristol Institution in the setting up of a separate Philosophical and Literary Society where local savants might give papers in a more intimate atmosphere. However, the Phil. and Lit. was housed in the same building, membership of it was open to all members of the Institution, without election, and in the period 1823–60 almost all members of the BI were members of its active satellite, the Phil. and Lit.

Bristol saw then an identity between those who financed its philanthropic and scientific activities; within the Bristol Institution, there was identity between members of the BI and the Phil. and Lit.; and only then does a real distinction appear, between the managers of these societies, all of whom were eminent civic figures, and scientific activists such as Prichard and Conybeare.

One central figure, neither cleric nor scientist, but a leading art patron, requires highlighting, since he was the leading medium through whom the BI contacted the Oxbridge and metropolitan conservative culture which it imitated and to which it sought attachment. This was John Scandrett Harford (1785–1866), FRS, inheritor of the brass and banking fortunes of that family, a Quaker convert to Anglicanism,[20] and a man who gave up mercantile pursuits to orchestrate the growth of Tory activities in Bristol. He was close to Hannah More, who gave financial support to the BI and whose obituary he wrote; a correspondent of Whewell's at Cambridge, a friend of Wilberforce, and an activist in almost all the evangelical societies in the city. A large landowner in Cardiganshire, Harford published in 1818 an attack on the life and character of Tom Paine.[21] He was a central organizational force in the BI throughout its active life.

The BI was divided then into two sections, with overlapping membership.[22] There was however one important structural absence in its organization: attempts to amalgamate with the Bristol Library in King Street came to nothing. The non-existence of a library of any great size contributed to the brief life of the BI.

The question then arises as to the role of local medical men in the Institution, and more importantly in the Phil. and Lit. Although they played an important role, they were not dominant intellectually until at least the mid 1840s, when the context of local scientific culture had changed decisively, partly because of developments within the medical profession itself. J. C. Prichard, Andrew Carrick, George Wallis, Henry Riley, J. E. Stock, the Unitarian eye surgeon J. B. Estlin, W. B. Carpenter, J. A. Symonds – these medical figures played an important intellectual role. But the BI was as much the home of the quirky antiquarianism of the Reverend John Skinner (1772–1839) as it was the province of local medical men.[23] The constituency of the BI was wide, taking in members from South Wales, including the Dowlais ironmaster J. J. Guest (1785–1852), the Dowlais trustee and engineer and archaeologist G. T. Clark (1809–98) and the Monmouthshire landowner and Poor Law reformer J. H. Moggridge (1771–1834).

The Bristol Institution was eventually built on a difficult site near the cathedral at the cost of nearly £11,500, with C. R. Cockerell (1788–1863) as architect. Cockerell's neo-classical style was then in vogue, as the most recent version of ancient aesthetic perfection. The building was financed by the purchasing of shares, with the families mentioned earlier purchasing the bulk of them. Cockerell's other work, in Bristol and elsewhere, on commerical and scientific buildings, illustrates in architectural terms the social origins of the BI.[24] The official roster of the BI included automatic membership for the city's two MPs, and of course a series of Presidents drawn from the local aristocracy: Lord Grenville, President until 1833, the Duke of Beaufort, and the Marquis of Lansdowne.

Examining the lectures and activities of the Institution first of all, we can see an active lecture programme for the years 1823–36, in a setting as deliberately patrician as possible. An unusual "special lecture" given by Norton Webster to a screened audience of mechanics in December 1824 was continually singled out in the BI records as an example of munificence and of correct social behaviour on the part of the "invitees". The social relations of attendance upon scientific occasions were explicitly hierarchical:

In endeavouring to render an Institution designed for and supported by the higher classes, in any degree subservient to such purposes, the acquisition of knowledge is represented in its most beneficial and becoming relation, as a boon, emanating from the superior to the inferior; and this, far from exerting any disorganising influence, becomes a new and strong bond in the fabric of society.[25]

It was precisely through a culture that surveyed scientific questions, ranging from religion to statistics, that the revived language of hierarchy would be found, and a collaboration within the urban elite effected.[26]

The pattern of lectures at the BI up to 1836 was varied, and to try to see an integrated ideology within them may be dangerous. But the lecturers were pursued and invited who deliberately evoked the tone and grandeur of the higher reaches of the "progressive" sciences. Charles Daubeny (1795–1867) newly appointed Professor of chemistry at Oxford, gave an opening address at the BI in January 1823, assuring his audience of the strengths that science (chemistry especially) might derive from commerce; his visit was important because of local family relations. In a nice twist of pretension and pathos, the BI advertised chemistry lectures, also in 1823, by 'Mr. Davy'; in fact the lecture committee had only managed to hire Edmund Davy (1785–1857), Sir Humphry's cousin, and it was always regrettable to them that the one-time Bristol inhabitant Sir Humphry had slipped through their fingers. Other popular lectures were those of Sir J. E. Smith (1759–1828), the purchaser of the Linnean Collection (1825); those of the tamed Jacobin John Thelwall, talking about Shakespeare and, in a double course of 1829, on elocution and reading; those of the Unitarian Lant Carpenter on the 'powers of the mind' (1830); and, despite little press cover, the phrenology lectures of Spurzheim (1827). On the literary side, the 'Marquis Spineto of the Royal Institution' gave popular lectures on literature in 1826.

The BI was however notably short of instruments, and had almost no utilitarian pay offs from its scientific activities. It housed very little apparatus, containing only a lucernal microscope, an air pump, some metereological instruments, and 'Attwood's beautiful machine for estimating vertical motion'. When Thomas Webster, secretary of the Geological Society of London, lectured on geology in 1828, the BI activist George Cumberland started a newspaper correspondence in which local landowners were accused of ignoring the useful information that Webster had presented to them as clues to places to dig for coal.[27]

A striking feature of other BI lectures up to 1836 (and indeed beyond) – lectures on natural history, zoology, on the eye, on the distribution of flora and fauna – was the insistent repetition of "creationist" philosophy. The argument from Design, the miraculous creation of species, the perfection of vertebrate structure, all are reiterated and glossed, with references to Paley, Derham and the Christian Cuvierisme of de Blainville. The Christian version of history and nature was reinforced by the art exhibition held at the BI: these combined the scenes of local pastoralism painted by Bristol artists, with the addition of "old masters" from the Harford and Miles collections, and the millennial paintings of Biblical epics. By far the most popular of these was Francis Danby's 'The Opening of the Seventh Seal'.[28]

While the lectures at the BI may have been diffuse, the activities of the Phil. and Lit. were more integrated. Here J. C. Prichard and W. D. Conybeare were dominant, with Conybeare especially maintaining links with other provincial and metropolitan savants, for example Vernon Harcourt in York. One way of extending links was through the election of honorary members, and these included Agassiz, Blumenbach, Lyell, Cuvier, Stanislaus Grottanelli of Siena, Marc Auguste Pictet, the Reverend John MacHenry [*sic*] and Benjamin Silliman.[29]

Conybeare, the Christ Church geologist and eventual Dean of Llandaff, worked hard for his Phil. and Lit. even when he left Brislington for Sully, Glamorganshire, in 1827. He had lectured on the history of scientific institutions, and had of course lectured on geology. His friend H. T. De la Beche (1796–1855) played a small part in the early history of the Phil. and Lit. and delivered a paper on 'The Diluvium of Jamaica' in May 1825. In good Bristol fashion, De la Beche was at the time in transit, from being a slave owner to becoming a scientist.[30] Conybeare's most striking didactic success was probably his scientific explanation for (and donation of) the *icthyosaurus communis*, which along with his work on the *plesiosaurus*, contributed to his ideas on successive creations of 'linked' marine reptiles, based on Cuvier's catastrophism and thus a firm answer to Lamarckian transmutation.[31] Likewise, Cony-beare's politics were important, stressing the need for moderate Whigs to hold together with conservatives against radicals or obscurantist, anti-progressive Tories. This idea of the common programme for alert conservatives is, I wish to suggest, an important factor in looking at the social uses of early nineteenth-century science.

The political differences between Conybeare and Prichard were not as important as their common scientific commitments. Prichard's work was far too influential to summarize here, but his influence in Bristol was great.[32] Prichard voted Tory or for the more conservative candidates, whenever he could, and identified himself completely with conservative activities in science and social life. The achievement of his ethnology, now discussed as part of 'the history of anthropology', in fact requires a different emphasis. He had produced a complete explanatory framework for the history of the "civilizing process", and for the movement of man from a primitive, black form toward the white heights of decorum embodied in the evangelical Anglican conservatism that he constantly defended. Religions that tended toward polytheism and animal worship, centrally that of the Egyptians, were examined at the Phil. and Lit., by Prichard among others, as examples of degeneracy. His ethnology supported an interventionist programme of Christianization, of the kind represented by the charitable and scientific groups at work in Bristol, because the existence of these elite activities was taken as being itself indicative of moral progress.[33] In terms of the history of science Prichard's most influential paper was probably the paper of early 1824, read in two parts, 'On the Distribution of Plants and Animals'. It was certainly made use of, when printed in a different form, by Lyell and Darwin. But as a social ideologist, employing analogies of "domestication" and sexual selection, his work was the scientific proof of the degradation of contemporary non-Christian peoples both abroad and in the metropolis.[34]

The BI and its Phil. and Lit. provided the Bristol upper class with an arena for a Christian conservative science that transcended denominational differences. The museum of the BI for example had various exhibits that contributed to this framework of a "progressive" creationist philosophy, including a gift from Buckland and Conybeare of bones from the Kirkdale cave; a series of rocks collected at Freiberg by Richard Bright (the younger) under the supervision of Werner himself, and an extensive collection of

British birds, together with the icthyosaur. Bristol, with its art exhibitions and reading-room, had at last answered the cultural challenge laid down by that model of the cultured merchant, William Roscoe of Liverpool.[35] The Institution's members awaited the arrival of the British Association in 1836 with bated but confident breath.

But the BI was not the only example of scientific culture in the city in the years 1820–60. The local Mechanics' Institute, founded in June 1825, was designed to extend the hierarchic imagery of nature and educational progress into the city's petty-bourgeoisie. Its financial base was similar to that of the BI, and derived from the same individuals who started the higher institution. One slight difference between the two was that Unitarians tended to deliver a higher proportion of the lectures at the Mechanics' Institute. But, despite this, Stephen Cave (1764–1838) a Tory Anglican banker was Treasurer, and Conybeare Vice President.[36] Subscribers were warned that the Institute needed to set a good example, to assuage the doubts accompanying the Mechanics' Institute movement in some quarters. Charles Pinney (1793–1867), West India merchant, Anglican, Mayor of Bristol at the time of the riots of 1831, and activist in the paternalistic General Trades Association and the Lord's Day Observance Society, as well as the Bristol Institution, stated at a MI meeting in November 1825 that

the high marks towards which their talents were to be directed was to trace the footsteps of the omniscient Creator throughout all Nature. . .in this way they would disarm the objections of their opponents and allure them to become promoters of that Institution which can thus make them better men and better citizens. . . .[37]

The MI had enrolled about 280 members by the end of 1825 and developed a small elementary school which taught mathematics. It tottered along until the early 1840s, with donations of books from such figures as H. H. Wilson (1786–1860), Professor of Sanscrit at Oxford; Samuel Lee (1783–1852), who received a stall at Bristol Cathedral in 1831, became vicar of Banwell in the Mendips, site of a famous bone cave, and then Professor of Arabic at Oxford in that same year; and J. A. Cramer (1793–1848), Thomas Arnold's successor as Regius Professor of modern history at Oxford from 1842. It is not surprising that the "mechanics" of Bristol did not find gifts from such abstruse sources of immediate delight. The MI continued with its reading room and library, and lectures from a

variety of figures such as the Chevalier Mascarenhas, the Portu-
guese Consul, in 1837; a 'Mr Cantor', described as a 'learned
foreigner' lectured on anthropology, and, in October 1844, 'Dr
Owens, MRCS and Mr W. J. Vernon of London' lectured on
mesmerism. But this typical example of an unsuccessful educative
institution disappeared in an amalgamation with the Clergy Book
Society in 1845 to form the Bristol Athenaeum.[38]

The Athenaeum was in fact the home already adopted by yet
another scientific society with a more noticeable utilitarian orientation, the little known Society of Enquirers.[39] This began in 1823,
almost in defiance of the BI, and met weekly in the Masonic Hall in
Broad Street. Lectures were given by local men almost exclusively,
although contacts were made with the metropolis in the form of the
Metereological Society of London. The activities of the society
were sympathetically covered by the Liberal *Bristol Mercury*, and
its two chief activists were the chemist and botanist Samuel
Rootsey, FLS, and the philosophical chemist William Herapath (d.
1868). Both these men found the BI 'exclusive'; and Herapath in
1828 accused the Institution of 'entertaining science, not of
promoting it'. And when a member of the Enquirers argued that no
expense be spared in the propogation of religion – 'Science', he
said, 'was less important' – the committee forced his resignation.[40]
Herapath was a well known local radical, active in the Bristol
Political Union in the Reform period; and he and Rootsey (who had
at times been imprisoned for debt) raised the issue of the chemical
state of the river Frome, effectively an open sewer in the city centre.
Interestingly, both Herapath and Rootsey came to hold positions in
the Bristol Medical School (1832), Herapath as lecturer in chemical
toxicology (1832–67) and Rootsey as botanical lecturer (1832–54).
The Society of Enquirers was important as the foundation of a
genuinely local, petty-bourgeois culture distinct from the BI, and
presaged the independent spirit that the organized medical profession was to bring.[41]

No extended popular scientific culture can be traced in Bristol in
these years. Phrenology did not appear as an organized movement
(a Mr G. Burgess and a Joseph Marriott read heads in the Arcade
around 1840) and perhaps the only two popular scientific debates of
the time were those of Robert Owen defending socialism in
December 1840 in a "Hall of Science" in Broadmead, and the
debate between the American elocutionist Jonathan Barber and
John Brindley, a native of Chester, on phrenology, in 1842.[42]

Developing in parallel to the career of the chemists of the Medical School was one other provincial career that throws some light on the relations between metropolis and province: that of museum curator. The curators at the BI included a German *émigré*, J. S. Miller (d. 1830) who had done distinguished work on crinoidea, had tried to get a job at the British Museum in succession to W. E. Leach, but was defeated by J. G. Children (1777–1852). Conybeare engineered his post at the Bristol Institution in 1823 at a salary of £150 a year. Miller, whose work was acknowledged by Blumenbach, Cuvier and Buckland, never completed a second scientific project on fossilized corals, and worked long hours at the BI cataloguing the "gifts" of its patrons. The English provinces did not bring particular success to this original and unrecognized scientist,[43] who died prematurely from these labours. His successors developed wider scientific links. Samuel Stutchbury (1798–1859), who had worked at the museum of the College of Surgeons, travelled in the Pacific as a naturalist, and became well known in Bristol for his museum and coal surveying activities. He also developed contacts (as Miller had done) with Oxford, through P. B. Duncan (1772–1863), a scientific activist in Bath as well as keeper of the Ashmolean Museum. Stutchbury attended Lyell at the inquiry into the Haswell Colliery explosion in 1844, extended the use of BI museum specimens in the lecture courses on comparative zoology, and eventually returned to Australia in 1850 to undertake a survey of New South Wales.[44] The career of museum curator received its ultimate metropolitan accolade – state support – when Robert Etheridge, curator at the BI from 1851 to 1856, became one of the curators at the Museum of Practical Geology in London.[45] The new interest of the state in scientific affairs was also obvious in the career of the Bristol geologist William Sanders (1799–1875), whom De la Beche recruited to supply information for the health of towns commission in 1844–5. Sanders looked after the museum of the Institution at various times in its history.

Meanwhile the BI itself was ailing. Before its demise, the Bristol Institution and its membership had three last rites to perform before the imitation of, and penetration into, the conservative culture it replicated was complete. First, there was the visit of the British Association. This visit from the lions of science came only just in time. From 1833 onwards the BI was badly in debt, but the need to be the first non-university city to be visited (after York) remained paramount. That visit came in 1836, complete with Murchison and

Sedgwick attacking De la Beche over the geology of Devon, and with tours to the great commercial sights of the city for the savants. They gazed at the piers of the proposed Clifton suspension bridge, even though this remained incomplete until the 1864 visit of the same Association to Bath. In a nice example of local dependence on metropolitan materials, the bridge was finished by using discarded parts from London's Hungerford bridge.[46] After the BAAS visit, no printed annual report for the BI appeared for a decade, and its financial crisis worsened. The only utilitarian project launched by the BI, the keeping of tidal readings, petered out after two years. Not even a visit to the Phil. and Lit. by the project's instigator, William Whewell, in July 1838, could sustain the efforts of his local tide-reading agent, the surveyor T. G. Bunt. Likewise, the only magazine to have been generated by the BI activists, the *West of England Journal* of 1835–6, with articles by Conybeare, Prichard, Stutchbury and the physician J. A. Symonds, perished.[47]

The BAAS visit helped to generate one other scientific development: the growth of a local statistical society.[48] Organized by members of the ailing Institution, the activists included the historian Henry Hallam (1777–1859), who had family connections with nearby Clevedon Court, Somerset; the conservative Anglicans J. C. Prichard, J. M. Gutch and the Reverend J. E. Bromby, the merchant and American consul Harman Visger (1802–67), the attorney George Bengough (1793–1856), and the society's chief activist, the Anglican soap manufacturer Charles Bowles Fripp (1806–49). The Unitarian minister Lant Carpenter participated, as did the physician J. A. Symonds. Taking the membership as a whole, the statistical society again represented the Bristol mixture of Anglican domination without exclusion of other elite religious persuasions. The purpose was simple: that of uncovering the state of educational and moral backwardness among the Bristol proletariat, and reporting on it. A particular cause for concern was 'sansculottism', 'which. . .exists in its miserable abode, and is ready, at any season of public weakness and agitation, to sally forth to its work of destruction. . .'.[49] Once the statistical findings had been published, the society folded (1842). But at least it had been part of the national statistical movement, with which Hallam, its chairman, was of course firmly associated.[50]

One last act of establishment was effected by the activists of the BI. This was the setting up of a non-denominational Bristol College which lasted from 1830 to 1841. Again, the composition of its activist core overlaps with that of the BI (with the possible proviso that Lant Carpenter was less involved, but then he ran his own educational academy): hence it was not a "godless college', with liberal inclinations, that fell victim to the religious objections of the influential evangelical T. T. Biddulph (1763–1838), and of two Bishops of Bristol, Robert Gray (1762–1834) and James Henry Monk (1784-1856). In fact, the college curriculum embodied precisely that Oxbridge mixture of classics and mathematics, with linguistic and theological accompaniments, that dominated the established educational institutions of the time. The objections of Gray (which were mild) and of Monk (famous for his extreme and idiosyncratic conservatism)[51] should not make one assume the presence of a centre of utilitarian, Benthamite godlessness. The purpose was simply to replicate Winchester, Eton and West-minster, and shepherd the sons of the college proprietors into Oxford and Cambridge. Conybeare acted as College examiner, and its staff included J. H. Jerrard, late lecturer in classics and Fellow of Gonville and Caius, Cambridge; the future physiologist W. B. Carpenter (1813–85), son of Lant Carpenter, and F. W. Newman (1805–97), tutor in classics and brother of J. H. Newman.[52] Success attended most of the sons entered into the College, the Prichards and Waytes not least, its most famous pupil being the economist Walter Bagehot.

The demise of the College and the collapse of the statistical society accompanied a deepening of the financial crisis in the BI. In 1837 its debts amounted to £1410, and they remained at over £500 throughout the 1840s. Membership fell from 223 in 1846 to 142 in 1860. Various enforced democratizations, especially in opening the museum to a wider public, did lead to a 'series' of popular lectures. The most influential of these were those of W. B. Carpenter which provided a useful platform for his general statements on zoology. They also allowed Carpenter to prepare the ground for his move to London and his career at University College.[53] Attempts to rescue the BI by other methods were not successful: a proposed amalga-mation with a newly formed Society of Arts, in 1852, came to nothing.[54]

This period of slack in the cultural activity of Bristol was compensated for by the medical profession. With the opening of the Bristol General Hospital in 1832, along with the Bristol Medical School, the founding of the Bristol branch of the Provincial Medical and Surgical Association in 1840, medical men became dominant in local scientific life. Two figures in particular, Henry Riley (1797–1848) and J.A. Symonds, both physicians, figured prominently in the activities of the ailing Institution. Having failed to achieve professorial status at the Phil. and Lit., Prichard proposed the foundation of a 'medical university' in Bristol in 1840.[55] But the relative independence of the medical community in Bristol is emphasized by the fact that Bristol College and the medical school did *not* amalgamate at this time. In a review of the history of the BI from 1851 to 1861, Symonds noted the increased presence of the medical profession among those who gave lectures.[56] But the inroad made into the Phil. and Lit. by the medical men was a shared one: the meetings of the society were also used by a group of local naturalists, some of whom, such as the botanist G. H. K. Thwaites (1811–82) also worked in the medical school, and it was this group that formed the centre of the Bristol Naturalists Society in 1861.[57] The development of specialized coteries of interested individuals (with overlaps) was completed by the founding of the Bristol Microscopical Society in 1843. Members included Thwaites, W. B. Carpenter, the eye surgeon J. B. Estlin, and William Budd.[58] Certainly the scientific initiative had moved to the medical men of the city, out of the decayed form of the Bristol Institution.

In this period of lapse, between 1840 and 1860, other forms of cultural activity did not cease, but were pursued in a fragmented context: for instance, there were visits from the National Association for the Promotion of Social Science, and the Archaeological Institute (1851), occurring alongside the modified culture of the doctors and the naturalists. Mary Carpenter conducted her social rescuing from Prichard's old residence, the Red Lodge. The arrival of the railway, the reorganization of the Corporation, the reform of the Corporation of the Poor in the 1850s – these among others were indices of local alterations in the mid century. With the establishment of Clifton College in 1860, and the amalgamation of the Bristol Institution and the Bristol Library, completed in 1871, the "Liberal Culture" had fully arrived. Samuel Morley was one of Bristol's Liberal MPs, and the Liberal families of Wills and Fry became synonymous with the financing of cultural and educational institutions.

The commercial city of Bristol had therefore generated a scientific culture that was, for most of the nineteenth century, only partially successful. Even at the end of the century, the local University College received little enthusiastic financial support in its early stages. Both the economist Alfred Marshall, then in Bristol, and his wife, deemed science and commerce to be poor partners, and the Bristol well-to-do slow to come forward with funds. Marshall was anxious to return to Cambridge, 'the model of unattainable perfection'.[59] And the existence of the Bristol Institution had not meant the production of a series of practical benefits for Bristol and its neighbourhood. It is striking for example that the first genuinely detailed geological map of the Bristol area was produced by William Sanders but not completed until 1862. The BI had not generated utilitarian aids for the port or local manufactories.

Perhaps commerce and science were not natural partners. But despite its institutional transcience, all the more evident when one compares the history of the BI with that, say, of the Leeds Philosophical and Literary Society, the culture of Bristol in these years remains of interest.[60] For Bristol had made an ideological connection with the established culture of Oxford and Cambridge, and done so through the mediating influence of science and scientific contacts. An urban, commercial elite, embarrassed by charges of philistinism, had developed a polite culture where natural theology, statistical research and the enforcement of Christian accounts of history, ethnology and geology became a common language. The scientific foundations of this design were the common origins of men and the intricate handiwork of the heavens. And this "structure of feeling" reflected the particular social preponderance of charities, philanthropic activities and exclusive government in the city's other institutions.

Bristol mimicked and thereby attached itself to the "progressive" culture of non-industrial England, and its elite had made their contribution to the scientific dimensions of Victorian conservatism. The culture of imitation, that stretched from Cambridge, through certain London societies, certainly through Oxford, now added Bristol to its constituency. Even in the twentieth century, this network continued to work, despite some alterations in the financial bases of cultural and educational institutions. Between 1820 and 1860, metropolis and non-industrial province had effected an alliance that remains visible.

Acknowledgements

For their help of various kinds, I would like to thank Bill Bynum, Roger Cooter, John Crump, Jan Dalley, Roy Porter, W. Watts

Miller and Mary-Kay Wilmers, and the staffs of the Bristol Corporation Archives Office, the Central Reference Library, the Museum and Art Gallery. For permission to cite material in its care, I thank the Bristol Corporation Archives Office.

Notes and references

1   A. W. Thackray, 'Natural knowledge in cultural context: the Manchester model', *American Historical Review*, 1974, **79**, 672–709; S. Shapin, 'The Pottery Philosophical Society, 1819–35: an examination of the cultural uses of provincial science', *Science Studies*, 1972, **2**, 311–36; S. Shapin 'Property, patronage and the politics of science: the founding of the Royal Society of Edinburgh', *British Journal for the History of Science*, 1974, **7**, 1–41. See also J. B. Morrell, 'Individualism and the structure of British science in 1830', *Historical Studies in the Physical Sciences*, 1971, **3**, 183–204; I. Inkster, 'Science and society in the metropolis: a preliminary examination of the social and economic context of the Askesian Society of London, 1796–1807', *Annals of Science*, 1977, **34**, 1–32; I. Inkster, 'Culture, institutions and urbanity: the itinerant science lecturer in Sheffield 1790–1850', in S. Pollard and C. Holmes (eds.), *Essays in the Economic and Social History of South Yorkshire*, Sheffield, 1976, 218–32; J. N. Hays 'Science in the city: the London Institution 1819–40', *British Journal for the History of Science*, 1974, **7**, 146–62.

2   M. Berman, *Social Change and Scientific Organization: The Royal Institution 1799–1844*, London, 1978.

3   S. F. Cannon, *Science in Culture: The Early Victorian Period*, Folkestone, 1978.

4   E. P. Thompson, 'The peculiarities of the English', in his *The Poverty of Theory*, London, 1978; D. C. Moore, *The Politics of Deference*, Hassocks, Sussex, 1976; particularly J. R. Vincent, *The Formation of the British Liberal Party*, 2nd edn, Hassocks, Sussex, 1976. See also J. E. Cookson, *Lord Liverpool's Administration*, Edinburgh, 1975, and G. K. Clark, *Churchmen and the Condition of England, 1832–85*, London, 1973.

5   For this idea in a different setting see D. Outram 'The language of natural power: the "éloges" of Georges Cuvier', *History of Science*, 1978, **16**, 153–78.

6   A. Hume, *The Learned Societies and Printing Clubs of the United Kingdom*, London, 1847; S. Shapin and A. Thackray, 'Prosopography as a research tool in history of science: the British scientific

community 1700–1900', *History of Science*, 1974, **11**, 1–28. Bristol does not figure prominently in R. M. MacLeod, J. R. Friday and C. Gregor, *The Corresponding Societies of the British Association, 1883–1939*, London, 1975.

7   For the history of Bristol, see J. Latimer, *Annals of Bristol in the Nineteenth Century*, Bristol, 1887 (reprinted 1970); P. McGrath and J. Cannon (eds.), *Essays in Bristol and Gloucestershire History*, Bristol, 1976; B. Little, *The City and County of Bristol*, London, 1954; A. B. Beaven, *Bristol Lists: Municipal and Miscellaneous*, Bristol, 1899; S. Hutton, *Bristol and its Famous Associations*, Bristol, 1907; V. A. Eyles, 'Scientific activities in the Bristol region in the past', in C. A. MacInnes and W. F. Whittard (eds.), *Bristol and its Adjoining Counties*, Bristol, 1955 (little on Bristol Institution); J. F. Nicholls and J. Taylor, *Bristol Past and Present*, i, Bristol, 1881; C. Tovey, *Bristol City Library: Its Founders and Benefactors*, Bristol, 1853; D. J. Carter, 'The social and political influences of the Bristol churches, 1830–1914', MLitt. thesis, University of Bristol, 1971; G. W. A. Bush, *Bristol and its Municipal Government, 1820–1851*, Bristol, 1976; H. E. Meller, *Leisure and the Changing City, 1870–1914*, London, 1976.

8   Catalogue for the exhibition, 'Science and music in eighteenth century Bath', Bath, September–December 1977. For the view that eighteenth-century Bristol lacked cultural institutions, see J. Latimer, *Annals of Bristol in the Eighteenth Century*, Bristol, 1887 (reprinted 1970), pp. 527–8 and *The New Bristol Guide*, Bristol, 1801.

9   For this periodization and the insights it offers on the "Liberal Cultural" renaissance from 1870 see Meller, op. cit. (7). Here one is talking of polite culture, and not Bristol's thriving popular culture.

10 I do not see "Liberal Anglicanism" as "progressive", because of the accompanying interest in science. Instead I see the "Liberal Anglican" movement as an awakened and attentive conservatism, which saw itself as having nothing to fear from science or social investigation. See Cannon, op. cit. (3), 1–71; cf. D. Forbes, *The Liberal Anglican Idea of History*, Cambridge, 1952.

11 W. Minchinton, 'Bristol – metropolis of the West in the eighteenth century', *Transactions of the Royal Historical Society*, 5th series, **4**, 69–89.

12 For Bristol's nineteenth-century economic decline, B. Alford, 'The economic development of Bristol in the nineteenth century; an enigma?' in McGrath and Cannon, op. cit. (7) pp. 252–83.

13   For the Corporation, Bush, op. cit. (7) pp. 17–41.

14   Alford, op. cit. (12), p. 265. Two firms which did flourish and survive in the second half of the century, Wills and Frys, developed outside this "commercial freemasonry" to form the basis of the "Liberal Culture" of the later period.

15   Information culled from the Tory newspaper, *Felix Farley's Bristol Journal* (henceforth *FFBJ*).

16   *FFBJ*, 4 February 1809. Other notable advertised lectures included those of S. T. Coleridge, speaking on Milton at the White Lion in Broad Street, *FFBJ*, 9 April 1814; and John Thelwall (1764–1834) who lectured on Milton and Shakespeare in June 1817, one among many visits by Thelwall.

17   J. E. Stock, *Life of Thomas Beddoes*, London and Bristol, 1811, pp. 145–7; G. Munro Smith, *A History of the Bristol Infirmary*, Bristol, 1917, pp. 369–70. On Beddoes, F. F. Cartwright, *The English Pioneers of Anaesthesia*, Bristol and London, 1952.

18   See *FFBJ*, 5 October 1822. For evangelicism, social reform and the Church, I. Bradley, *The Call to Seriousness*, London, 1976; E. R. Norman, *Church and Society in England, 1770–1970*, Oxford, 1976; and F. K. Brown, *Fathers of the Victorians: The Age of Wilberforce*, Cambridge, 1961.

19   Despite John Kaye (1783–1853), Bishop of Bristol 1820–7, the activists in the science of the city were invariably the Deans. Beeke's successor was John Lamb (1789–1850), a Whig Anglican, Master of Corpus Christi College, Cambridge, from 1822. Kaye was Regius Professor of divinity at Cambridge from 1816 and 'an able, deeply conservative High Church prelate'; R. A. Soloway, *Prelates and People*, London, 1969, p. 237. Beeke himself wrote on the question of income tax, defending its 'moral equity'.

20   A movement into conservatism that he shared with Prichard, who was enabled to attend the University of Oxford.

21   Harford was also involved in the setting up of St David's College, Lampeter.

22   My account of the internal history of the BI is based on: records of the BI and the Phil. and Lit., Archives Office of the Corporation of Bristol, no. 32079, items 1–152 (includes minute books, registration of proprietors, letter books, cash books and correspondence); Jefferies Collection of Bristol Central Library, College Green, Bristol, especially vols. 1 and 2, numbered B 26066/26065; miscellaneous materials in Bristol Central Library, including printed annual reports of the BI for 1823–36 (B 4508–4521); Bristol Museum

and Art Gallery, for printed reports for 1846 and 1860. These two sporadic publications indicate the troubled history of the BI from 1836 onwards.

23   Biographical information in G. Munro Smith, op. cit. (17). The BI purchased fossil collections, but did not purchase a collection of anatomical drawings by Pauli Mascagni, as suggested by the Siena Professor of medicine, Grottanelli. On medical men in a different context, I. Inkster, 'Marginal men: aspects of the social role of the medical community in Sheffield 1790–1850', in J. Woodward and D. Richards (eds.), *Health Care and Popular Medicine in Nineteenth Century England*, London, 1977, pp. 128–63. For Skinner, H. and P. Coombs (eds.), *Journal of a Somerset Rector*, Bath, 1971.

24   D. Watkin, *The Life and Work of C. R. Cockerell*, London, 1974. Cockerell worked on projects such as banks as well as the new Ashmolean Museum at Oxford. Shares in the BI cost £25 each, which was more than one of the Institution's "assistants", a relative of the religious fanatic Joanna Southcott, received per annum.

25   BI second annual report, p. 25.

26   Meller, op. cit. (7), p. 48, suggests that

The activities of the Institution had provided the basis for the integration of social groups within the city formerly without the means, or possibly the desire, for intercommunication. By 1861, the process had not gone all that far. But the framework for future integration among the city's upper classes had been laid.

27   For Cumberland's letters, part of which appeared also to impugn the geology of W. D. Conybeare and W. Phillips's *Outline of the Geology of England and Wales*, London, 1822, see *FFBJ*, 13 December 1828, p. 3. On Cumberland, G. E. Bentley, Jnr, *A Bibliography of George Cumberland*, New York and London, 1975.

28   T. Fawcett, *The Rise of English Provincial Art*, Oxford, 1974, esp. pp. 132–3 and 186–7 on Bristol. On the "Bristol School", F. Greenacre, *The Bristol School of Artists: Francis Danby and Painting in Bristol 1810–1840*, Bristol, 1973.

29   A list of honorary members up to 1836 is given in the annual report for each year.

30   P. J. McCartney, *Henry De la Beche: Observations on an Observer*, Cardiff, 1977.

31   Conybeare donated an icthyosaurus to the Museum of the Institution; his later lectures to the Phil. and Lit., for example on the landslip at Lyme Regis of 1824 (delivered 28 April 1825) were also evidence of the exposure of fossil remains through local, "cata-

strophic" action. See *FFBJ*, 27 May 1825, p. 4. On Conybeare, M. J. S. Rudwick, *The Meaning of Fossils*, London, 1972, p. 144, and F. J. North, 'Dean Conybeare, geologist', in *Transactions of the Cardiff Naturalists Society*, 1933, **66**, 15–68.

32   I owe much of my information on Prichard to John Crump, but all the interpretations are my own.

33   Prichard gave papers on mummies (January 1825), on ethnology at various times, on the history of pestilences (October 1827), on the doctrine of a vital principle (November 1828), on philology (April 1832), and on phrenology and animal magnetism (January 1835). This is an incomplete list, but indicates that he used the society as a mouthpiece.

34   For anthropology, G. W. Stocking Jnr, 'From chronology to ethnology: James Cowles Prichard and British anthropology, 1800–1850', in Prichard, *Researches into the Physical History of Man*, London, 1813, ed. G. W. Stocking, Chicago, 1973, pp. ix–cx; and G. Weber 'Science and society in nineteenth century anthropology', *History of Science*, 1974, **12**, 260–83.

35   William Roscoe (1753–1831), first President of the Liverpool Royal Institution, was always cited as the example to be followed by the Bristol merchant classes.

36   *FFBJ*, 25 June 1825, p. 3.

37   *FFBJ*, 5 November 1825. The budget for the MI for the year 1825–6 was a mere £165.

38   Latimer, op. cit. (7) pp. 288–9. For Mechanics' Institutes as failed attempts at bourgeois persuasion, see S. Shapin and B. Barnes, 'Science, nature and control: interpreting Mechanics' Institutes', *Social Studies of Science*, 1977, **7**, 31–74, and I. Inkster, 'The social context of an educational movement: a revisionist approach to the English mechanics' institutes, 1820–1850', *Oxford Review of Education*, 1976, **2**, 277–307.

39   Almost all the evidence comes from newspapers; no printed membership lists appear to exist.

40   See 'Proceedings of the Society of Enquirers: a reply to E. V. Rippingille', Bristol pamphlets, B2309, Bristol Central Library.

41   Lectures at the Society of Enquirers, when on natural history (for instance by James Prowse, February 1827) were Prichardian, monogenist and Christian. References to natural theology abounded in other lectures. The chemists provided the distinctive utilitarian dimension. On Herapath and Rootsey, A. Prichard, *The Early History of the Bristol Medical School*, Bristol, 1892, pp. 7–11.

42 Brindley challenged both Owen and Barber, and was assisted *contra* Owen by John Scandrett Harford. The "Hall of Science" became a Liberal Party meeting place from 1843: Latimer, op. cit. (7), p. 251. Marriott was trained as a lawyer, but turned to dissenting religion and popular science. Activities such as mesmerism were often followed by elite figures, most importantly the second Earl of Ducie (1802–53).

43 On Miller (or Müller), W. D. Conybeare, *The Philosophical Magazine or Annals of Philosophy*, January 1831, pp. 3–7; and 'H. J.', *Bath and Bristol Magazine*, 1832–4, pp. 111–22. J. G. Children was appointed to the BM post because of his affiliation with Humphry Davy, the "lost scientist" of Bristol.

44 For Stutchbury, see the annual and museum reports for the BI, 1831–6, and the minute books of the museum sub-committee 32079 [20] and [21]; Curators' reports are in the Jefferies Collection, B 26066, on the BI, for 1827–57.

45 On Etheridge, *Geological Magazine*, 1904, **5**, 1.

46 Latimer, op. cit. (7), p. 375.

47 This magazine was edited by G. T. Clark.

48 For a discussion of the statistical movement in general, but also one where the information regarding Bristol is inaccurate, see M. J. Cullen, *The Statistical Movement in Early Victorian Britain*, Hassocks, Sussex, 1975, pp. 121–3. P. Abrams, *The Origins of British Sociology, 1834–1914*, Chicago, 1968, pp. 35–8, is ingenious. For the reports of the Statistical Society in Bristol from September 1836–40, vol. B 4592, Bristol Central Library, which omits the last meeting of 1842. Honorary members of the Bristol Statistical Society included the Glasgow statistician James Cleland, and W. R. Greg, (1809–81) who had been educated in Bristol by Lant Carpenter.

49 First and second annual reports of the society.

50 An agent hired from the Central Society of Education had travelled through the districts of Temple, St Michael and St James, counting heads, occupations and Bibles. C. B. Fripp, who had presented statistical information on Bristol to the 1836 BAAS, also produced a later report on the state of the Bristol working class. Annual reports also gave details about police returns, types of crime committed annually, and attendance at Sunday schools.

51 See Soloway, op. cit. (19), pp. 249–52 and 302–3.

52 On the origins of Bristol College, *FFBJ*, 28 November 1829 and 20 August 1830. For Gutch and the Tory *Journal*, the College formed part of the conservative cultural revival of the city, along with the Institution and the building of new churches.

53 For the part played by these courses in the development of Carpenters's career, and their place in the drawing up of his physiology textbooks, J. E. Carpenter, 'Introductory memoir', in W. B. Carpenter, *Nature and Man*, London, 1888, pp. 27–9.

54 'Address of the Committee of the Bristol Institution', drawn up by S. S. Wayte, January 1848, Bristol Archives Office, 32079 [82].

55 *FFBJ*, 20 June 1840. In his study of the Yorkshire Philosophical Society, A. D. Orange notes similar links between the YPS and the York Medical School; see his *Philosophers and Provincials: The Yorkshire Philosophical Society from 1822 to 1844*, York, 1973.

56 J. A. Symonds, *Ten Years*, Bristol, 1861. The "prosopography" is: gentlemen of no profession, 2; gentlemen of commerce, 3; gentlemen of the profession of literature, 3; architects, 3; gentlemen of science, 5; of law 7; of education, 8; of the clerical profession, 13; of the medical profession, 19.

57 D. E. Allen, *The Naturalist in Britain*, London, 1976, ch. 8.

58 M. Pelling, *Cholera, Fever and English Medicine*, Oxford, 1978, pp. 156–63. She slightly exaggerates the role of Dissent in amalgamating the medical community in Bristol.

59 M. Sanderson, *The Universities and British Industry, 1850–1970*, London, 1972, p. 70.

60 See E. K. Clark, *The History of 100 Years of Life of the Leeds Philosophical and Literary Society*, Leeds, 1924, pp. 234–5, where membership figures increase markedly in the years that mark the imminent demise of the BI.

7 Rational dissent and provincial science: William Turner and the Newcastle Literary and Philosophical Society

Derek Orange

Introduction

A few yards from John Dobson's imposing Central Station in Newcastle upon Tyne stands the Literary and Philosophical Society, founded by William Turner in 1793, accommodated in its present home since 1825. Today it sponsors an occasional celebrity lecture, hosts a variety of musical activities, and supports a small local history group. But for most practical purposes the Society is a private library, struggling like others of its vanishing kind to maintain a welcoming oasis of reading and conversation in the wilderness of the office blocks. The books may be dusty but the coffee is excellent.

The intellectual life of a provincial town during the early decades of the industrial revolution, however, was not (or not for long) comprehended within the walls of a single institution. Next door to the Lit. and Phil., on a site occupied in the 1850s by the Newcastle College of Medicine and now the headquarters of the North of England Institute of Mining Engineers, the Society of Antiquaries of Newcastle upon Tyne, founded in 1813, conducts its monthly business, although its official home is in the Black Gate of the Castle and its pre-Conquest collections are now in keeping of the University. The University is also responsible, together with the Natural History Society of Northumberland, Durham and Newcastle upon Tyne, established in 1829, for the plenitudinous Hancock Museum. The antiquarians and the naturalists were once part of the Lit. and Phil., the President of the College of Medicine was *its* President and both the Mining Institute and the University properly number it among their ancestors.[1]

Tucked away between the Central Station and the Castle is Hanover Square, an area of dingy offices and small workshops. Before the coming of the railway, the Square boasted some of the best houses in the town, as well as the Unitarian chapel where

William Turner ministered for almost sixty years. The present Unitarian congregation, unremarkable and unremarked, meets a mile away in Ellison Place. Their angular twentieth-century building announces itself, proudly enough, as the Church of the Divine Unity, and the home of liberal Christianity in Newcastle since 1662. From time to time it displays a small poster which proposes the pertinent but logically impenetrable question, 'Are you a Unitarian without knowing it?'

This essay is concerned with the Newcastle Literary and Philosophical Society (henceforth LPS) during the first forty years of its existence, with William Turner and, to a degree, with the religious springs of his intellectual mission. It is an exercise not so much in Thackray's 'cultural geography' of science as in its topography. The budding of provincial scientific and educational initiative during the pubescence of industrial Britain is a phenomenon of absorbing interest. But the institutions which were the expression of that initiative deserve to be looked at first in their own terms. That they appeared and flourished in different places at different times and with differences of emphasis and effectiveness should discourage premature or facile generalization. Here, although the magnification is variable, perhaps dangerously variable, the field of view is designedly narrow.

In 1830, then, Newcastle could boast, in addition to a Mechanics' Institute, three societies which were, in some sense, scientific. The LPS, the largest of them, numbered about 800 members. In 1838 there were several hundred local members of the British Association for the Advancement of Science at its meeting in the town. Clearly "community support" for science was markedly more extensive than the size of the "scientific community" (however it is characterized) would indicate.[2] What, then, was the character of the oldest of the societies? What, precisely, *were* its "literary" and "philosophical" pretensions? How were they to be attained, and by whom? In what degree were they, in fact, realized? And if, inevitably, there were failures, were they attributable to indifference, inexperience or incompetence, or to hesitations and diversions brought about by competing conceptions and purposes? Whatever the answer to these questions, the historian of Newcastle science must come to terms with two facts: that for almost half a century William Turner was a consistent and continuing factor in its evolution, a scarcely less familiar figure among the Antiquarians, say, or in the Mechanics' Institute, than in the LPS; and that the members of his congregation exerted a disproportionately large influence on the direction and course of Novocastrian institutions.[3]

Since a conjunction of this kind can hardly be dismissed as fortuitous, it is appropriate to ask what light it sheds upon the ever interesting problem of the relation of liberal Christianity (or rational dissent as Turner more naturally called it) to the spread of science and education in the England of George III. Does the commonly offered analysis in terms of the civil disabilities, alternative value-systems and emasculated theology of the Unitarians prove adequate, or do local and personal variables need to be identified? What, in particular, did the juxtaposed responsibilities of a dissenting pastorate and the management of a provincial scientific society mean for Turner?

II

The local context

In a letter written in 1801 Turner could still refer to Newcastle as 'a country-town'.[4] With, perhaps, 30,000 inhabitants (and more than 50,000 in 1831),[5] it was nevertheless the most considerable centre of population between Leeds and York to the south and Edinburgh to the north. Its history, stretching back to Roman times, was largely the history of its river (in the 1820s statistically it was the second seaport in the kingdom) and, more recently, of its mineral resources. The great northern families spread their estates through the counties of Durham and Northumberland, particularly along the valleys of the Tyne and its tributaries to the west of the town, their wealth and position based on land, on coal, and on lead. On the lower reaches of the river there was iron-smelting, and soap- and glass-making, and the beginnings of a chemical industry; and (although not yet on the daunting scale of the age of steam and iron) there was shipbuilding and ship-repairing.

Given this situation, the country town was also a regional capital: the Newcastle and Northumberland Assizes attained an importance as much social as legal; the Infirmary was opened in 1752, the Dispensary in 1777; the "new" Assembly Rooms of 1776, with their attendant facilities for cards and coffee and newspapers were held locally to rank next to those of Bath; the Theatre Royal, erected in 1788 at the beginning of an era of civic development, flourished under the management of Stephen Kemble; there were regular subscription concerts and occasional musical festivals; the popularity of the midsummer Race Week warranted the erection of an elegant stone grandstand on the Town Moor.[6]

The buoyancy of the intellectual life was the measure of a level of population and prosperity, and of confidence, which the French wars did nothing to diminish. There was a certain symbolism in the demolition of sections of the old town walls at the turn of the century: not until the vanquishing of the Jacobites in 1746 was Newcastle finally delivered from the insecurity of life on the frontier.[7] But the physical separateness from London, and the emotional separateness from everywhere south of the Tees or north of the Tweed remained. The natives of Tyneside, then as now, combined a fierce pride in their homeland with an open contempt for those dark powers by whose caprice its excellence went unrecognized.

The LPS was heir to a number of intellectual traditions in Newcastle. Probably the earliest source of scientific enlightenment was the lecture course. The record, incomplete as it is, extends from the early years of the eighteenth century when James Jurin, briefly Master of the Royal Free Grammar School in the town (and later Secretary of the Royal Society of London), was reputed to have amassed £1000 by his lectures on experimental philosophy, through the exertions of John Horsley of Morpeth, and Caleb and John Rotheram, and Isaac Thomson and Robert Harrison, to James Ferguson's discourses in Charles Hutton's highly patronized school in 1770.[8]

A Philosophical Society was active in Newcastle for perhaps three or four years from 1775. The period of the American war was a sensitive one ('the best of times. . .the worst of times') and the philosophers were quick to defend themselves against the objections 'that meetings of this kind . . . become schools of sedition and infidelity'. More dramatically, they expelled the young radical Thomas Spence for printing and circulating copies of a paper which he had read to the Society on 'The real Rights of Man'. By the turn of the century Spence had become a figure of some notoriety in London, and the LPS was obliged several times to disclaim any connection with the earlier body. That there was no close kinship between the two institutions was true enough. But it was well known in Newcastle that some of the most venerated members of the LPS had, in their youth, been associated with the Philosophical Society. At least two of them, Robert Doubleday and William

Chapman, still survived at the time of Spence's death in 1821.[9]

It may be doubted whether the Philosophical Society of the 1770s was ever more than a debating club whose score of members were youthfully preoccupied with the social and political issues of the day. The Philosophical and Medical Society (henceforth PMS), which came into being in 1786 and followed a rather sterile existence until its dissolution in 1800, assumed a more dignified stance. The opening of the Infirmary and the Dispensary, as well as the prospect of a lucrative general practice, brought to Newcastle men of high ability, many of them (like John Rotheram the first President of the Society) physicians with degrees from Edinburgh, to compose something of an intellectual elite. The Gentlemen of the Faculty were not, however, equally accomplished in the conduct of a society; and their meetings, largely devoted to the reading and discussion of case histories, became more irregular and more disorganized. In the early 1790s there was an extraordinary spate of resignations, several of those who departed having held presidential or vice presidential office.

A dozen or more of the forty members of the PMS attached themselves to the LPS in 1793, most of them bearing office in both institutions during the seven or eight years of coexistence. The annals of the older Society contain a pregnant sentence which apparently came to light too late to exercise the imaginations of the nineteenth-century local historians. In May 1789 'the Secretary presented a Card from the Rev. Mr. Turner, requesting a sight of Dr. Rotheram's introductory paper read at the institution of the Society, which request was complied with, on condition that Mr. Turner does not take a copy of it, nor suffer any other person to read it'.[10]

In 1787 the PMS posted notices in the Assembly Rooms and the most frequented coffee-houses ('Katy's' and 'Bella's') soliciting support for the establishment of a general library devoted to 'the most approved Authors in every Branch of Philosophy: in the Belles Lettres, History, Theology, Law and Medicine, Agriculture and Commerce'. They also attempted, with an equal lack of response, to secure the library in St Nicholas's Churchyard as a meeting-place. The library, instituted in mid century under the bequest of Dr Robert Tomlinson, remained uncatalogued and neglected. Its appointed custodians were as reluctant to facilitate public access when Eneas MacKenzie published his history of Newcastle in 1827 as they had been when the bookseller William

Charnley had made representations to the Bishop of Durham in 1789. Charnley himself had opened a circulating library in 1757; at the beginning of the nineteenth century, now managed by Robert Sands, it comprised about 8000 volumes and was acclaimed as 'one of the best collections of books, in every branch of science and literature, out of the metropolis'.[11] Libraries, like debating societies, were regarded with official suspicion at the time: the Corresponding Societies Act of 1799 rendered it illegal to maintain either a circulating library or a reading room except under licence granted annually by two magistrates.

III

The Newcastle Literary and Philosophical Society: foundation and development

Whether or not the conception of the LPS owed anything to the sons of Hippocrates, its birth took place without the benefits of their attendance. William Turner was one of a small circle of friends (James Clephan dignified it with the name, club) who met weekly for intellectual conversation. On one of these occasions, towards the end of 1792 the group met, on the outskirts of the town, in the home of Robert Page the controller of customs in Newcastle. Turner pointed out, apparently casually enough, the need for a local society for the reading of papers and discussion on literary and scientific subjects. He was immediately requested by Page and Malin Sorsbie, merchant and accountant, to draw up a statement of his case. Turner's pamphlet, under the cautious title 'Speculations on the Propriety of Attempting the Establishment of a Literary Society in Newcastle' was circulated for some weeks and discussed in a preparatory way before what was now a 'Literary and Philosophical Society' was formally proposed at a meeting in the Assembly Rooms on 24 January 1793, and instituted at a second gathering on 7 February in the Dispensary. Turner, elected as joint secretary, addressed to the first regular meeting on 12 March a series of 'Further Observations and Hints on the leading objects of the Society; and on the conduct of its members'. This statement, closely related to the 'Speculations', provided the substance of the subsequently printed *Plan of the Literary and Philosophical Society of Newcastle upon Tyne*.[12]

If the event was unpremeditated, the man was not unprepared.

After ten years in Newcastle, Turner, as his *Address* to the members of his congregation in 1792 reveals, was engaged in serious self-examination.[13] His chapel and the school which he had started were now fixed points in his life; but the petitioning of Parliament for the civil rehabilitation of dissenters, a work in which he had been vigorously engaged, had now come to an unwelcome hiatus. While the wider issues raised by the engagement of Protestant dissenters in science and education must be reserved for a later part of this essay, it is easy to discern in Turner's earlier life circumstances which predisposed him to the cultivation of intellectual activities in Newcastle. His sojourn at Warrington Academy, where the tradition of rational inquiry and of free discussion between tutor and student was sedulously cultivated, had ended in 1781, in which year the Manchester Literary and Philosophical Society had come into formal existence. Turner, acquainted through his Warrington friends with many of the Mancunian philosophers, had read a paper to the Society in November 1783 and had been elected an honorary member.[14] But if the Manchester Society had attained a size and a formality which he did not immediately contemplate in Newcastle, he could look to the example of the Lunar Society in Birmingham and Joseph Priestley, the now persecuted friend of his father, or to the Derby Philosophical Society fashioned by his distant kinsman Erasmus Darwin.

In recent years scholars including Schofield, Musson and Robinson, and Porter have commented on the marked preoccupation of the Newcastle LPS with applied science and technology.[15] Two observations may be made. The first (which can be illustrated immediately) is that the conviction that a provincial society should espouse strongly utilitarian, even industrial interests, is one that Turner derived from Priestley and recognized (or thought he recognized) in the activities of the Manchester Society. The second observation (which requires a larger portion of the essay to elaborate) is that the programme proposed in the foundation documents of an institution did not necessarily or for very long correspond to the course actually followed.

Turner's *Speculations* naturally enough, then, echoed some of the arguments and even the phrases of the Preface to the 1785 volume of *Manchester Memoirs*. He commended Manchester for diverging from the tradition by which, in England, philosophical societies were generally confined to the metropolis, and for creating an institution which had led several members to 'pursuits connected with the improvement of its extensive manufactures'.[16] In 'Further Observations and Hints' he rehearsed explicitly the argument which Priestley had advanced: while metropolitan bodies attended to the general interests of science, it was in the investigation and improvement of their local situations that provincial societies would find their proper business as well as their best chance of attaining 'utility, reputation and success'.[17]

Turner seems to have employed the term 'Literary Society' for two reasons. First, to distinguish his proposed institution from the PMS whose members properly confined their attention to the practical aspects of their profession. Second, a 'Literary' society denoted, at any rate for Turner, a society of a certain character: a group, probably small in number, which received verbal and written communications and undertook to consider them with calmness and good humour, 'the investigation of truth', as he declared in a Priestleyan phrase, 'being no further interesting to any person, than as truth may be found thereby'.[18]

Whatever Turner's intentions as to the constituency and the conduct of the Society, his *Speculations* left readers in no doubt that its concern was primarily with natural knowledge. It was not that the literary and visual arts were excluded from its interests: if the institution was to address itself comprehensively to the local situation they could not be. It was rather that they were deliberately subordinated: lead, coal, and the manufactures which depended on coal, iron and steel, glass and pottery and chemicals, were to be the near points of the Society's gaze. Moreover, it was a worthy object to consider how far the physical environment as a whole was still (the word is very characteristic of the man and the period) *improvable*. The Society might reassess the mineral resources of the region, act as an exchange for agricultural intelligence, promote the augmenting of the water supply, undertake the 'enumeration' and 'classification' of the inhabitants of the town.[19]

There is plenty of evidence that during the earliest years the members of the Society saw themselves as implementing the programme which Turner had proposed. The papers read to the monthly meetings during the first five years showed a strong predisposition to topics related to geology, mineralogy, and their applications, to industrial (and especially chemical) processes, and to agriculture. Of particular interest are the proposals set before the Society by William Thomas in September 1796, 'Hints for the formation of a plan to be proposed to the Coal Owners for establishing an office in Newcastle for recording various important information respecting the Coal-Works and Wastes in the neighbourhood'; the account of the reduction of lead from its ore and its separation from silver by James Mulcaster, Director of the works at Langley Mill, Northumberland, for the Governors and Commissioners of the Greenwich Hospital; Joseph Garnett's telegraph of 1795 (which Turner commended, without success, to the Admiralty); and Turner's own 'Sketch of the Commerce and Manufactures of Newcastle' in 1802. Turner was quickly associated through his personal friends with the British Mineralogical Society and with a Friendly Association of Ironmasters of Yorkshire and Derbyshire. There were many exchanges of compliments and information between the LPS and the new and ineffectual Board of Agriculture and its chairman Sir John Sinclair.[20]

In retrospect, the long-term realization of Turner's programme became less probable when the Society took to itself that other ambition close to the hearts of scholarly Novocastrians, the establishment of a general library. In the very early days of *Speculations*, Turner had sought the support of the Reverend Edward Moises, Master of the Royal Free Grammar School in the town. At the tenth monthly meeting of the LPS in December 1793, the reading of a letter from Moises to Turner resulted in the setting up of a committee 'to consider the outlines of a Plan for carrying into more immediate effect the establishment of a General Library'.[21]

Turner dated the beginning of the Society's library from the meeting, such a provision, as he contended, not having formed part of the original plan of the Society. There had certainly been no suggestion of the kind in his *Speculations*. Nevertheless, the constitution adopted at the second of the meetings (with Moises in the chair) which marked the formal institution of the Society had left it to the future deliberations of the members 'to determine what, or whether any, measures should be taken for obtaining the establishment of a general library'. History was thus at odds with prehistory.

By 1807 the LPS library contained about 8000 volumes. Throughout the period of his secretaryship Turner viewed its growth and popularity with some misgiving. A colloquium which consisted of a small number of active members meeting regularly for discussion and exchange of scientific intelligence was one thing; an institution the majority of whose members regarded the library as its *raison d'être*, neglected its monthly meetings and appeared at its annual meetings, if at all, to advance the interests of the library against its wider concerns, was quite another.

To study the history of the Newcastle LPS is thus to be confronted by an institution which in its earliest years conformed closely enough to the intentions of its founder, but which in a comparatively short time relegated those matters to the periphery of its activities, dismissed them as special interests to be pursued by groups of enthusiasts inside, or better outside, the institution. Turner had to look for the unfolding of his programme not in the LPS but in the Newcastle Society of Antiquaries, and in the Natural History Society of Northumberland, Durham and Newcastle upon Tyne. In each case, although the men concerned typically continued as members of the original body, it was the new society which commanded both their organizational thrust and their philosophical communications. That the library of the LPS quickly revealed itself as the cuckoo which ousted the parental eggs is a situation of which Robert Spence Watson was well aware, although his conviction is obscured by the expansive and anecdotal method of his centenary history. But the assertion had already been made emphatically enough by R. M. Glover in the jubilee year of the institution: 'our Society has now become little more than a large reading club'.[22]

The tensions between a reading club and a scientific society were perhaps first publicly recognized and debated following the establishment of the so-called "New Institution" in 1802. If there was no anticipation of a library in Turner's *Speculations*, neither was there any explicit proposal for organized lectures. Nevertheless the LPS facilitated, accommodated, or sponsored courses by Dr Henry Moyes, by the fantastical Dr Katterfelto, by George Wilkinson the Sunderland surgeon, by T. O. Warwick of Rotherham, and by John Stancliffe of the Middlesex Hospital. Most of the lectures clustered around the years 1798–1801, and a paper read in December 1798 in which Turner urged 'the introduction of Courses of Lectures on subjects connected with the happiness of mankind as members of Society'. The immediate outcome of the paper was an invitation to Dr Thomas Garnett, Professor of natural philosophy and chemistry in Anderson's Institution in Glasgow, to provide such a course during the summer of 1799. Garnett accepted the commission, but the death of his wife and his subsequent removal to the Royal Institution in London prevented its fulfilment.[23]

The case for a more regular provision of lectures was put forward by Thomas Bigge, one of the local gentry and a Vice President of the

LPS in a paper; 'On the Expediency of establishing a lectureship in Newcastle on subjects of natural and experimental Philosophy' which he read in May 1802 and afterwards summarized in an *Address to the Public*. Bigge's advocacy embodied an unequivocally utilitarian view of science. He emphasized the Baconian cadence of his phrase 'the New Institution' with a passage from *The Advancement of Learning* which commended 'such natural philosophy as shall not vanish in the fumes of subtile, sublime, or delectable speculation, but such as shall be *operative in the endowment and benefit of man's life*'.[24] For Bigge the point of science, at least in the local situation, lay in the exploitation of natural resources: 'Without mechanics and chemistry, how wretched would be our manufactures, and how worthless the amount of our mineral possessions!' The very prosperity of their region, he added in the *Address*, was 'founded upon the principles of natural and experimental philosophy'.[25]

William Turner was quickly (some said too quickly) appointed Lecturer to the New Institution. He prepared himself by acquiring Thomas Garnett's lecture notes, which he did not use, and John Rotheram's apparatus which, with some lack of dexterity, he did. His *Introductory Discourse*, delivered and published in the following November, echoed many of Bigge's themes even as Bigge had echoed his own. The Newcastle region, he argued, illustrated 'the absolute dependence of. . .Commerce upon science, and the necessary connection between the extension of the one and the enlightened application of the other'. Chemistry and natural philosophy were indispensable in a seat of arts and manufactures and a great seaport. While Bigge found his inspiration in the newly formed Royal Institution, Turner looked to the recent lectures of Dr John Alderson (an honorary member of the LPS) in Hull, as well as to those of William Farish at Cambridge.[26]

IV

Debate and division

Turner himself first revealed the existence of divisions in the Society concerning the New Institution. In August 1808 he addressed the members in a *Letter*, written in response to a communication he had received which, over the name "Mentor" criticized the Lectureship and claimed to voice 'the opinions of the Members of the Society at large' in doing so. "Mentor" soon turned out to be Ralph Beilby, one of the earliest supporters of the LPS, a committee member, and a man by no means unfriendly to Turner.[27]

Turner hoped that the publication of his *Letter* would encourage 'a full, free and candid discussion' of the controversial issue and, no doubt, of the larger purposes of the Society. In fact it provoked an exchange of rhetoric and raillery. Three stormy meetings in the autumn saw the support of the Society for the New Institution reaffirmed in the face of hostile speeches by Beilby, Moises, John Marshall the Librarian, and others. A winter of pamphleteering was marked by the resignation of Beilby from the committee and Moises from the Society. A conciliatory dinner was organized in April 1809 and the lectures continued – mechanics, electricity and magnetism, chemistry, optics, astronomy and occasionally natural history – until Turner retired from his office in 1833.[28]

Whatever dissatisfaction there was in the Society had a longer history than was immediately apparent. The opponents of the New Institution lectureship claimed that it had not been generally popular among the members since its introduction in 1802. Turner in his reply to "Mentor" admitted that since his appointment he had received anonymous addresses often enough to convince him that there were people connected with the Society who were not friendly 'either to the Institution or the Lecturer'.[29] The earliest of several surviving squibs is dated 1803. Moreover the *Historical Sketch* of the Society, written in 1807, almost certainly by Turner, and a manuscript probably utilized in its composition, 'Extracts from the Minutes to show that the Library has not been the Original or Exclusive Object in the Expenditure of the Society's Funds',[30] clearly embodied a particular (and to the critics a partial) view of the LPS and its objects.

Although there were those who saw the New Institution as having brought new members and new resources to the Society and those who emphatically denied the claim, more fundamental issues were at stake. The *Historical Sketch* traced the establishment of the library and the consequent redeployment of most of the funds of the Society to the second year of its existence; and pointed to the decision of the committee in 1795 to support a repository for objects of natural history, the interest in the following spring in the contemplated purchase of a mineralogical collection from Dresden, the unconsummated liaison with Garnett and so on. But Beilby had urged the pressing necessity 'to secure the great interests of the

Society upon its original and proper foundation – an extensive, valuable, and well-chosen LIBRARY'. Those who sympathized with him either reiterated that the New Institution was a departure from the original objects of the Society *or*, that if narrowly philosophical activities such as lectures and collections had been counted among the early objects, they had quickly proved unpopular and unworkable. At a more practical level, the delivery of the New Institution lectures imposed limitations upon the use of the premises and particularly the much patronized library. Hence Turner's not unsuccessful attempt, in 1809, to defuse the situation by finding alternative accommodation for the lectures.

While there were indeed practical and personal aspects, it is illuminating to look at the controversy as uncovering different notions as to what a provincial Lit. and Phil. might do and might be. First of all, it can be seen as a division of opinion on the social function of the Society. Turner's unflagging commitment to education as an instrument of social and moral improvement is manifest in his oversight of the Charity Schools attached to his chapel, in his early advocacy of Sunday Schools, in his central role in the establishment and conduct of the Royal Jubilee Schools in Newcastle. That Bigge shared a similar concern is illustrated most interestingly in his management of *The Œconomist or Englishman's Magazine*, a relatively short-lived monthly publication to which Turner was an occasional contributor.[31] For both men, the New Institution offered a possible means of extending beyond the level of the favoured middle class the benefits of scientific knowledge. Bigge looked to 'the miner, the mechanic, the manufacturer, and the agriculturalist', Turner to 'the merchant, the manufacturer, the miner, the engineer, the shipwright, and the navigator'. They found one of their most influential supporters in James Losh. Losh, a close personal friend of Bigge and a contributor to his magazine, had been associated as a young man with Thomas Beddoes in the introduction of Sunday Schools and Schools of Industry in Bath; even in his old age he and Turner energetically supported Brougham's Society for the Diffusion of Useful Knowledge.[32]

If, then, the New Institution was seen by its originators and chief supporters as an expression and extension of the educational mission of the LPS, there can be no doubt that it was resisted by some members for the same reason. In 1824 Turner could, reasonably enough from his own point of view, see the new Mechanics' Institute as the offspring of the older society, while the radical Eneas Mackenzie could rumble darkly about the exclusiveness of 'the Dons' who (having raised the annual subscription to finance the erection of the new building) he found 'resolved to shut the doors ... in the face of all who do not have a heavy purse'.[33] George Stephenson, the most celebrated of Newcastle artisans, had received much encouragement from Turner, and from James Losh's brother, William, and on one famous occasion had demonstrated his "Geordie" miner's lamp to the LPS. But he was not a member of the Society.

The controversy of 1808 and 1809 must also be seen as setting lectureship against library, the "philosophical" allegiance of the Society against the "literary". Even after the restoration of peace, Turner perceived that defending a position was very different from winning a campaign. As long as the Society was treated by most of its members as no more than a library, the philosophical minority inevitably felt frustrated and unsupported. The contention in this essay is that the polarization of interests and priorities was responsible for the eventual creation of other institutions to promote the philosophical interest of Novocastrians.

In January 1814 Turner read to the monthly meeting of the LPS a paper (subsequently printed and circulated to the members) explaining the inauguration of a Literary Club. The Club consisted effectively of an inner society, a small group of members meeting monthly and undertaking to supply regular papers, the papers constituting a reservoir of material for the ordinary monthly meetings of the Society, whose normal supply had almost dried up: 'the number of contributors ... of late years, has born a sort of inverse ratio to the number of members, diminishing as the Society increased', Turner reported pointedly. The library, he went on, 'has become so much the principal object of attention, that the original purpose of the institution, that of meeting for literary discussion, has sometimes been in danger of being lost sight of'. The ordinary monthly meetings had inevitably declined 'to a mere nominal attendance, held (as some of our new members have supposed they were intended to be) for no other purpose than to ballot for proposed members'.[34]

A rough manuscript list of members of the Literary Club contains fifteen names and there is a comparable list in the diary of James Losh.[35] Significantly enough, Turner first referred to the group as 'the Literary *Society*': there is no doubt that he saw in it the possibility of creating or recreating the kind of philosophical circle which he had first envisaged twenty years earlier. In this he was destined to receive some encouragement. For the Literary Club continued to meet throughout the period covered by this essay and ranged widely in its concerns: archaeology, geology, education, political economy, astronomy and literature all appeared in its annals. So, too, did mining: it was before the Literary Club that William Chapman revived Thomas's proposal of 1796 to found a record office in Newcastle. But it was ironic that, although the Reverend John Hodgson, a member of the Club, had been instrumental in establishing a Society for Preventing Accidents in Coal Mines in 1813, and a number of LPS figures including Chapman and Turner sat on its committee, it was to *Sunderland* that the body which fathered the Davy safety-lamp owed its allegiance; in Newcastle it was the Society of Antiquaries, not the LPS, which marked the invention by electing Sir Humphry Davy to honorary membership; and when John Buddle, in his turn, urged the collection of mining records, it was the Natural History Society which he saw as its most appropriate repository.[36]

The Society of Antiquaries was in every sense a child of the LPS. For many years the two bodies shared not only premises, but President (Sir John Swinburne) and Secretary (John Adamson, joint secretary with Turner of the LPS; joint secretary of the SA from its foundation in 1813). In 1825 Turner reminded the LPS, assembled for the first time in its own building, that with the rise of the Antiquarians, 'one great object, originally contemplated' had been taken out of the hands of the parent body. He probably foresaw that the mood of the lay membership, hardened after the recent purchase of a comprehensive collection of specimens, would soon banish the naturalists as well. When, in 1829, the Natural History Society was duly formed, he attributed the 'culpable neglect of this essential part of our original plan' to 'the almost exclusive attention which for many years was paid to that confessedly important, but by no means exclusive object, the formation of a library'.[37]

V

Philosophers, professions and parties

In 1837, when Turner resigned his secretaryship, the LPS had surrendered to newer institutions many of the ambitions which it had professed in 1793. Nevertheless, in the forty years before steam-locomotion changed the character of Newcastle and its river, and industrialization began to appear threatening and disruptive, there was a high degree of stability in the leadership of the Society. Turner's perennial position was matched by that of Sir John Swinburne, President from 1798 to 1838. There were hardly more than a dozen holders of the vice presidential office during the same period, and perhaps 100 committee members. The safest generalization is that the effective direction of the LPS remained, as it had begun, in the hands of the professional classes and the local gentry. Swinburne might be an informed and popular President, but his family seat was at Capheaton, twenty miles away. The active vice presidents, and still more the committee members, were men who lived in Newcastle or its immediate precincts.[38]

That men who had been leaders of the PMS quickly found employment in the LPS has already been noticed. The next generation added a number of distinguished medical names: in particular T. E. Headlam, a Vice President of the Society from 1834 to 1850 and President from 1850 to 1855; and T. M. Greenhow, whose proposals for the establishment of a college or university in Newcastle, put before the Society in 1831, were regarded by R. M. Glover as the logical extension of the case expounded by Turner and Bigge, thirty years earlier.[39] The legal profession brought into the LPS men of high ability, among whom James Losh was perhaps the most powerful and John Adamson the most energetic. The third of the traditional professions, the Church, was, however, meagrely represented. After Moises, the only important clergymen of the Established Church were John Hodgson and Anthony Hedley, joint secretary before Adamson and archivist. Nor, apart from Turner, is it easy to detect any interest in the Society among dissenting ministers.

Moises was as much usher as cleric, and schoolmastering added several luminaries to brighten the third and fourth decades of the LPS, most notably John Bruce and Henry Atkinson.[40] The other professional group whose services both to the Society and to the cultural life of Newcastle were considerable consisted of booksellers, printers and newspaper proprietors: William and Emerson Charnley, booksellers; Solomon Hodgson and his two sons who conducted the independent *Newcastle Chronicle* and acted as literary midwives in the promotion of natural knowledge, from the works of Thomas Bewick to the reports of the local societies; John

Mitchell and W. A. Mitchell, proprietors of the radical *Tyne Mercury*; Eneas Mackenzie, radical printer, local historian and founder of the Mechanics' Institute; and John Marshall, equally radical, dismissed from his position as Librarian of the LPS in 1817 after publishing *A Political Litany*.[41]

The LPS sometimes looked askance at its radicals, but there can be no doubt that its mood was predominantly Whig during the period of the French wars and reformist after them. The room at the back of Charnley's bookshop which was for a couple of generations well known as the headquarters of the clique of Newcastle Whigs was also, by natural association, an informal rendezvous for the inner members of the LPS before the monthly meetings. The old Whig and Tory county families were, in fact, frequently brought together by marriage settlements and mining syndicates; nevertheless the Swinburnes and Ridleys lent their patronage, the Brandlings, the Carrs, the Ellisons did not. In the town, the influence of Edinburgh was strong: Losh and Headlam were on terms of some intimacy with Henry Brougham, while Hedley's sympathy with the politics of *The Edinburgh Review* was believed to have cost him ecclesiastical preferment. .

Non-conforming politics shaded not very subtly into non-conformist religion. James Losh often complained that his reputation as a Whig and a Dissenter damaged his professional prospects. In 1823 Henry Atkinson was denied a teaching post at the Grammar School by a decision which Turner condemned as 'more of religion and politics than mathematics'.[42] Not only was Atkinson's controversial support of the purchase of Byron's *Don Juan* by the LPS remembered, wrote W. A. Mitchell of the rejection, 'He was accused of being a radical! Then, of being a dissenter!'[43] The visible symbol of the marriage between liberal politics, liberal religion and useful knowledge was Turner's chapel in Hanover Square. Both Losh and Atkinson were members of the congregation. So, indeed, were very many of the active members of the LPS. They constituted, as one of the next generations of Newcastle historians recalled, 'a conspicuous constellation in the literary firmament of Tyneside'.[44]

VI

The dissenting connection

William Turner, the central character of the present essay, was born in 1761 and died in 1859. An instructive history of Protestant dissent in its most distinctive period might be fashioned out of the chronicles of his family. It would begin with his grandfather, John Turner (1689–1737), born in the year of the Revolution Settlement and the Toleration Act which offered a limited acknowledgement of the right of Dissenters to worship together, and portray him leading the men of his Preston congregation to protect the interests of the House of Hanover against the Old Pretender; it would show his anticipations of greater religious and civil liberty at the hands of the Whigs continually frustrated during the lifetime of his son, the first William Turner (1714–94); it would trace the evolution of dissenting theology from the strict Calvinism of John Turner through the residual Arianism of his son to the Unitarianism of his grandson; it would illustrate the parallel development of the academies, from Chorlton's household seminary in Manchester which John Turner attended, to Warrington, of which the second William Turner was first student and then historian, and Manchester (New) College, of which he was official Visitor.[45]

Such a history would necessarily lay bare the elaborate networks which marriage and inheritance, education and commerce, fashioned within the dissenting subculture. It would show the Turners united over more than one generation with the Hollands of Lancashire and Cheshire, and related to the Robberds and the Gaskells and the Martineaus, and the Wedgwoods and the Darwins. Nor would it be difficult to recognize in the history the operation of the dissenting connection. Turner's attempts to sustain the technological impulse which he had given to the LPS in Newcastle were supported by his contacts with the British Mineralogical Society through Arthur Aikin, the son of his most admired tutor at Warrington, and with the Friendly Society of Ironmasters through the Dawsons of Huddersfield and Bradford.[46] In 1811 Turner was in touch with Reverend Joseph Hunter, then in the early stage of his many years as a promoter of natural knowledge in Bath: Hunter's father, like Turner, had studied at Reverend Joseph Dawson's school in Idle, and Hunter himself at Manchester College, York, at the same time as Turner's son William.[47] In 1818, in the course of his visitorial address in Manchester College, Turner regretted the death of one of his correspondents, Dr John Thomson, a graduate of the College. Thomson, sometime President of the Royal Medical Society, was at the time of his death active in the promotion of a philosophical society in Leeds. Turner commended to his audience the encouragement of provincial institutions; the Principal of the College, Charles Wellbeloved, his son-in-law and senior tutor John Kenrick, and Turner's son William, were among the earliest members of the Yorkshire Philosophical Society founded in York in 1822. The junior William Turner was later secretary of the Literary and Philosophical Society in Halifax for many years.[48]

The most telling illustration of the continuity which William Turner perceived between rational dissent and provincial science must, however, be drawn from the critical early years of the LPS in Newcastle. In 1797 he was invited by the Trustees of Manchester College to undertake the direction of the institution. He declined the offer principally because, as he wrote to his brother, 'the Literary Society...is not yet fit to be left to itself; and I see no individual here who possesses the inclination, and at the same time the spirit and perseverance, to rear it to a state of maturity'.[49]

The social historian of science must always seek to achieve a proper balance between the general and the particular, to reflect both on culture and community and on the individual men and women who are at once creators and creatures of culture and community. Turner was a gentle and unpretentious man, whose transparent integrity and boundless charity were appreciated even by those who could not approve his religious unorthodoxy. Nor can his physical vigour and durabilty be disregarded: when the LPS came into being, Turner, though hardly more than 30, had ministered in Hanover Square for ten years; his school, the most highly regarded in Newcastle in the period between Hutton and the Bruces, gave him an added influence; by the turn of the century he was a Newcastle institution, to whom natives brought their problems and visitors their inquiries.[50]

But whatever Turner owed to nature or grace was multiplied in its effectiveness by those capacities which belonged in some degree to his kind: an entry into all levels of society; an intellectual base broad rather than deep, progressive rather than traditional; an acute sense of the value of truth (and therefore of education) as a liberating force in the lives of men; an alert social and political conscience. All these attributes were constantly confirmed and reinforced by his intercourse with intimates of like mind from whom he also received insights into what was going on in London and Edinburgh, and more especially in Manchester and Leeds and Birmingham and Boston.

In later as in earlier life, then, Turner was *Vigilii Filius*, a torch-bearer between dissenting generations. That he was happy to be identified with Unitarianism in its narrower sense is apparent from the energy with which he advertised the cause after its legalization in 1813; his establishment of a Unitarian Tract Society in Newcastle and his history of Warrington Academy in the pages of *The Monthly Repository* both belong to these years.[51] But, importantly, he resisted the application of the Unitarian name to his congregation:

this society was originally established, and has been all along maintained, upon the principle of the right of individual judgement in matters of religion. Its members, therefore, desire to be considered as a voluntary association – of individual Christians, each one professing Christianity for himself according to his own views of it, formed upon a mature consideration of the Scriptures, and acknowledging the minister's right to do the same; and necessarily united in nothing but a desire to worship the supreme Lord of all as the disciples of one common Master; and – to keep the unity of the spirit in the bond of peace, rather than the unity of faith in the bond of ignorance.[52]

Rational dissent was a legitimate child of the Reformation, and in this revealing profile some of its inherited characteristics are clearly visible: its scripturalism, its intellectualism, its individualism, its catholicity. The 'Sufficiency of Revelation', the title of one of Turner's published sermons, was a favourite theme of Unitarian preachers.[53] While Protestants generally brought to the Bible a reverence which passed easily into superstition and an understanding prejudiced by creed or confession, Unitarians came to the book with a gaze at once more innocent and more sophisticated, and declared themselves unable to discover in it the crucial doctrines of Trinity and Atonement. The view most widely held among them, Turner claimed in 1800, was that God was 'the merciful Parent of the Universe, who never meant anything but the happiness of his creatures'.[54] Jesus was indeed the common master, but prophet and teacher rather than redeemer, pointing those who repented of wickedness and showed in their lives the fruits of repentance to a heavenly bliss, and perhaps an earthly one. It was an optimistic view, consistent enough with the fashionable secular utopias of the time.
Within this 'multiform radical religious movement',[55] however,

there was no orthodoxy of content, only integrity of approach. Its genius and elusiveness lay in the freedom, indeed the responsibility, of individual inquiry which it commended both to minister and layman. The catholicity of liberal religion made it attractive to men of lofty views and high ambition. Whatever degree of social cohesion is attested by the networks of rational dissent, the tradition was far from a static one. Rising young men passed through Unitarianism on their way to the more respectable Anglicanism of a prosperous middle age. The chapel in Hanover Square had its long-established families – the Hodgsons, the Huttons, the Charnleys – whose members traced their ancestry to "the old dissent"; but the congregation was notable, even during the years when liberal opinions were in eclipse, for the number of first-generation Novocastrians and first-generation Unitarians it embraced: John Clark from Edinburgh, the Bruces from Alnwick, Pattinson and the Loshs from Cumberland.

The industrialization of Tyneside before 1840 was less harsh, less arbitrary than that of Manchester or Glasgow. But the growth and the new mobility of the population, the disintegration of the old town and the old boundaries, the visible expansion of traditional trades and the coming of new ones, and the prodigious busy-ness of the river did not fail to register in the consciousness of Novocastrians. That the values which Protestantism commended were, at an individual level, consonant with the evolution of a high capitalist society is well established. Rational dissent added to them a social dimension, a process theology which seemed to make sense of material change. The industrial revolution was to take place not, as it were, behind God's back but at his express command. In Turner's congregation the distinction between sacred and secular was obscure: the religious quest passed naturally into intellectual inquiry, the attempt to understand and to improve the physical environment was seen to be a religious duty. The kingdom was recognizably coming on earth, even as it was in heaven.

Acknowledgements

Unpublished archival material is cited by courtesy of the British Library, Newcastle City Libraries, the Archives Department of Tyne and Wear County Council, and the Literary and Philosophical Society of Newcastle Upon Tyne.

Notes and references

1 For Newcastle institutions, see R. S. Watson, *The History of the Literary and Philosophical Society of Newcastle upon Tyne (1793–1896)*, London, 1897, pp. 260–319; S. Middlebrook, *The Advancement of Knowledge in Newcastle upon Tyne: The Literary and Philosophical Society as an Educational Pioneer*, Newcastle upon Tyne, 1974.

2 R. H. Kargon, *Science in Victorian Manchester: Enterprise and Expertise*, Manchester, 1977, p. 1.

3 Obituaries of Turner: W. Gaskell, *Sermon on the Death of Rev. William Turner*, Manchester, 1859; J. Kenrick, *The Christian Reformer*, 1859, **15**, 351–66, 410–24, 454–61; J. Clephan, *The Gateshead Observer*, 30 April 1859, 7 May 1859. On the Hanover Square congregation: W. Turner, *A Short Sketch of the History of Protestant Nonconformity, and of the Society Assembling in Hanover Square, Newcastle*, Newcastle upon Tyne, 1811; R. Welford, *The Church and Congregation of the Divine Unity, Newcastle upon Tyne*, Newcastle upon Tyne, 1904; Unitarian Church records (microfilm), Tyne and Wear Archives Department.

4 Turner to *The Monthly Magazine*, 12 May 1801, **11**, 413.

5 In 1801 the census showed 28,294 inhabitants; in 1831, 53,613.

6 [J. Baillie], *An Impartial History of the Town and County of Newcastle upon Tyne*, Newcastle upon Tyne, 1801; [D. Akenhead], *The Picture of Newcastle upon Tyne*, Newcastle upon Tyne, 1807; 2nd edn [prepared by J. Hodgson], Newcastle upon Tyne, 1812; E. Mackenzie, *A Descriptive and Historical Account of the Town and County of Newcastle upon Tyne*, 2 vols. in one, Newcastle upon Tyne, 1827; the most considerable modern work is S. Middlebrook, *Newcastle upon Tyne: Its Growth and Achievement*, Newcastle upon Tyne and London, 1950; 2nd edn, Wakefield, 1968.

7 Middlebrook, op. cit. (1), p. 3.

8 J. Brand, *The History and Antiquities of the Town and County of the Town of Newcastle upon Tyne*, 2 vols., London, 1789. On Horsley: W. Turner, *Newcastle Magazine*, March 1821, **1**, 426–7; J. Hodgson, *Memoirs*, Newcastle upon Tyne, 1831. On Rotheram and his lectures: R. Welford, *Men of Mark 'twixt Tyne and Tweed*, 3 vols., London, 1895, **iii**, p. 328; J. Rotheram, *A Philosophical Essay on the Nature and Properties of Water*, Newcastle upon Tyne, 1770, p. 3; *DNB* confuses John Rotheram of Newcastle (1719–87) with one of his sons, also John Rotheram. On Ferguson's lectures: C. Hutton,

Tracts on Mathematical and Philosophical Subjects, 3 vols., London, 1812, **iii**, p. 379.

9 On Spence, the Philosophical Society and its members: O. D. Rudkin, *Thomas Spence and his Connections*, London, 1927, pp. 42–4; *Newcastle Magazine*, January–February 1821, **1**, 315; March 1821, **1**, 427–8.

10 D. Embleton, *History of the Medical School*, Newcastle upon Tyne, 1890, and *The Newcastle Medical Society a hundred years ago*, n.p., n.d.; G. G. Turner and W. D. Arnison, *The Newcastle upon Tyne School of Medicine, 1834–1934*, Newcastle upon Tyne, 1934; Mackenzie, op.cit.(6), pp. 501–27.

11 [Akenhead], op.cit. (6), p. 46; J. Knott, 'Circulating libraries in Newcastle in the eighteenth and nineteenth centuries', *Library History*, 1970–2, **2**, 227–49.

12 *Plan of the Literary and Philosophical Society of Newcastle upon Tyne, Instituted Feb. 7, 1793*, n.p., n.d. The chief collections of ephemera relating to the Newcastle LPS (with a great deal of duplication between them) are: 'Reports, Papers and Catalogues of the Literary and Philosophical Society of Newcastle upon Tyne', 11 vols., 1793–1824, collected by Anthony Hedley (henceforth RPC), Newcastle LPS library; and 'Literary and Philosophical Society MS papers', 12 vols., Newcastle City Library, local history collection. Other important sources in the Newcastle LPS library are: 'Papers read to the monthly meetings', 4 MS vols., 1793–1824; 'Literary and Philosophical Society, letters', 7 MS vols., 1793–1866.

13 W. Turner, *An Address to the Members of the Society of Protestant Dissenters in Hanover Square, Newcastle*, Newcastle upon Tyne, 1792.

14 Gaskell, op. cit. (3), pp. 10–11.

15 R. E. Schofield, *The Lunar Society of Birmingham: A Social History of Provincial Science and Industry in Eighteenth-Century England*, Oxford, 1963, p. 438; A. E. Musson and E. Robinson, *Science and Technology in the Industrial Revolution*, Manchester, 1969, p. 161; R. Porter, 'The industrial revolution and the rise of the science of geology', in M. Teich and R. Young (eds.), *Changing Perspectives in the History of Science*, London, 1973, pp. 320–43 (329–31).

16 W. Turner, *Speculations on the Propriety of Attempting the Establishment of a Literary Society in Newcastle*, Newcastle upon Tyne, 1793, p. 2.

17 Turner, 'Further Observations' in 'Papers read to the monthly meetings', **i**.

18   ibid., p. 7.
19   Turner, op. cit. (16), pp. 4–9.
20   'Papers read to the monthly meetings', **i**; RPC, **i** and **ii**.
21   RPC, **i**.
22   R. M. Glover, *Remarks on the History of the Literary and Philosophical Society of Newcastle upon Tyne*, Newcastle upon Tyne, 1844, p. 9. Glover's argument is similar to that put forward in the present essay.
23   RPC, **i**.
24   T. Bigge, *On the Expediency of Establishing a Lectureship in Newcastle on Subjects of Natural and Experimental Philosophy*, Newcastle upon Tyne, 1802, p. 1; Bigge, *Address to the Public*, Newcastle upon Tyne, 1802 (the Newcastle LPS contains a copy of the *Address* in a bound volume of tracts by Turner, which may indicate that Turner had at least a share in its composition).
25   Bigge, op.cit. (24), p. 7.
26   W. Turner, *General Introductory Discourse on the Objects, Advantages, and Intended Plan of the New Institution for Public Lectures on Natural Philosophy*, Newcastle upon Tyne, 1802.
27   W. Turner, *Letter to the Ordinary Members of the Literary and Philosophical Society of Newcastle*, Newcastle upon Tyne, 1808.
28   RPC, **iv**.
29   Turner, op.cit. (27), p. 3.
30   *A Historical Sketch of the Transactions of the Literary and Philosophical Society of Newcastle upon Tyne during the First Twelve Years from its Commencement*, Newcastle upon Tyne, 1807; 'Extracts from the Minutes to show that the Library has not been the Original or Exclusive Object in the Expenditure of the Society's Funds', RPC, **iv**.
31   *Œconomist, or Englishman's Magazine*, Newcastle upon Tyne, 1798 and 1799; W. Turner, *A Brief Memoir of the Late James Losh, Esq., Recorder of Newcastle*, Newcastle upon Tyne, 1833.
32   E. Hughes (ed.), *The Diaries and Correspondence of James Losh*, 2 vols., Durham and London, 1962 and 1963, **ii**, pp. 175–208; Brougham, with some errors of detail, had complimented Turner and the founders of the Mechanics' Institute in Newcastle in *Practical Observations upon the Education of the People*, London, 1825, p. 22.
33   Welford, op. cit. (8), **iii**, p. 117.
34   W. Turner, *Address to the Monthly Meeting of the Literary and Philosophical Society of Newcastle upon Tyne, Held January 4, 1814*, RPC, **v**.
35   Hughes, op.cit. (32), **i**, p. 30.

36   J. Raine, *A Memoir of the Rev. John Hodgson*, 2 vols., London, 1857, **i**, pp. 90–123, 170–89; RPC, **vi**; R. L. Galloway, *A History of Mining in Great Britain*, London, 1882, pp. 153–81: T. R. Goddard, *History of the Natural History Society of Northumberland, Durham and Newcastle upon Tyne*, Newcastle upon Tyne, 1929.

37   W. Turner, *Address Delivered at the First Meeting of the Literary and Philosophical Society of Newcastle upon Tyne Held in its New Apartments*, Newcastle upon Tyne, 1825, p. 16; Turner, *Introductory Address at the First Meeting of the Natural History Society of Northumberland, Durham and Newcastle upon Tyne*, Newcastle upon Tyne, 1829, p. 10.

38   Successive annual reports of the Society, bound in RPC.

39   Glover, op.cit. (22), pp. 18–19.

40   J. Bruce Williamson, *Memorials of John Bruce and Mary Bruce*, Newcastle upon Tyne, 1903; W. Turner, 'Memoir of the late Mr. Henry Atkinson', reprinted from *The Newcastle Magazine*, April 1829, **8**, 175–8.

41   On the Mitchells, Welford, op.cit. (8), **iii**, pp. 191–205; on Mackenzie, ibid., **iii**, pp. 114–19; on the Hodgsons, ibid., **ii**, pp. 543–55; on the dismissal of Marshall, RPC, **vii**.

42   Turner to A. Hedley, 14 January 1823, RPC, **xi**.

43   W. A. Mitchell, *Tyne Mercury*, 28 January 1823 and in RPC, **xi**.

44   Welford, op.cit. (8), **ii**, p. 455.

45   On the Turner family: Kenrick, op.cit. (3); W. Turner (III), *Lives of Eminent Unitarians*, 2 vols., London, 1840–3, **ii**, pp. 336–81. On Manchester College, V. D. Davis, *A History of Manchester College*, London, 1932. The College moved from Manchester to York in 1803, back to Manchester in 1840, and then via London settled in Oxford.

46   RPC, **ii**; *Historical Sketch*, op.cit. (30), p. xxii.

47   J. Hunter, 'William Turner, senior' (MS note, undated), BM 36527, fo. 54; Turner to Hunter, 12 October 1811, BM 24607, fo. 75.

48   W. Turner, *Address by the Visitor, Manchester College*, York, 1818. On the early members of the Yorkshire Philosophical Society, see A. D. Orange, *Philosophers and Provincials*, York, 1973.

49   Kenrick, op.cit. (3), 362.

50   ibid., pp. 410–13.

51   Turner's papers from *The Monthly Repository*, 1813–15, **8–10**, are conveniently reprinted, with an introduction by G. A. Carter, in W. Turner, *The Warrington Academy*, Warrington, 1957.

52   Turner, op.cit. (3), pp. 29–30.

53   W. Turner, *The Sufficiency of Revelation: A Sermon*, Newcastle upon Tyne, 1791.

54   W. Turner, *A Sermon Preached at the Chapel in Hanover Square, Newcastle, for the Support of the New College, Manchester*, Newcastle upon Tyne, 1800, pp. 9–10.

55   C. G. Bolam, J. Goring, H. L. Short and R. Thomas, *The English Presbyterians*, London, 1968, p. 233.

8 Economic and ornamental geology: the Geological and Polytechnic Society of the West Riding of Yorkshire, 1837–53

Jack Morrell

Recent work on institutionalized provincial science during the first English industrial revolution has suggested that, even in industrial areas, science was pursued and patronized more as polite learning and as a mode of cultural affirmation than as a direct agent of technical and economic change.[1] It has been likewise argued that during this period British mining contributed little to the creation of the science of geology and that geology aided mining even less.[2] In this essay I consider the extent to which such findings are illuminated by the early history of the Geological and Polytechnic Society of the West Riding of Yorkshire which was founded in 1837 by local coal-mine and iron-furnace owners and managers. At the same time I hope to add a few nuances to the detailed account in Davis's standard history of the Society by drawing on unpublished sources which were not available to him.[3]

I

Critics who scourged the state of English science in the early 1830s often singled out geology as the happy exception. Likewise the Geological Society of London (f. 1807) was depicted as a model scientific society for others in the metropolis to emulate. There is no doubt that compared with other disciplines geology was then extremely buoyant; and that the Geological Society tried to act as a metropolitan leader in English geology through its *Proceedings*, its *Transactions* and the famous discussions at its meetings. At the same time it was a gentleman's geological club dominated by a merito-cratic oligarchy based on scholarship and not on rank. Gentlemen of secure and in some cases fabulous income, such as Lyell, Murchison, Fitton, Darwin and Greenough, aided by Oxbridge clerical academics such as Sedgwick, Buckland and Whewell, dominated the Society: they led the manly discussions, dictated the social ethos, and shared the administrative spoils. With the obvious exception of John Taylor, the doyen of Cornish mining who was Treasurer from 1823 to 1843, surveyors, engineers, and mine owners were not key Fellows. Given the occupational composition of the Geological Society's oligarchy, its romantic wanderlust, its love of elevated and elevating scenery, it is not surprising that this metropolitan coterie generally neglected coal formations and mining areas.[4] In the 1830s the Geological Society of London nurtured the work on the Cambrian, Silurian and Devonian systems done by Sedgwick and Murchison, and on tertiary formations by Lyell and Fitton, without encouraging comparable work on either mining areas or the carboniferous coal measures.

Yorkshire geology had not been ignored by the metropolitan and Oxbridge geologists. In the early 1820s Buckland, Professor of geology at Oxford, had made Kirkdale cave and others famous for the fossil teeth and bones they contained. His sensational discoveries were, however, quite different from the sustained research programme pursued by Sedgwick, the Yorkshire-born Professor of geology at Cambridge. Sedgwick had worked on Yorkshire geology, but as a means and not an end. His study of Yorkshire magnesian limestone had attempted to establish its relation to what were subsequently called the Permian marls; and his research on the carboniferous Pennine chain of hills was undertaken not on its own account but primarily to illuminate the stratigraphical and dynamical geology of the Cumbrian mountains.[5]

This emphasis on polite Yorkshire geology was not confined to the Geological Society coterie. In the 1820s the Yorkshire Philosophical Society (f. 1822) had devoted itself to the geology of Yorkshire and the antiquities of York by collecting pertinent materials in its Museum at York. Taking advantage of both local opportunity and pride, the self-conscious provinciality of the Yorkshire Philosophical Society achieved rapid success: having appointed John Phillips as the Keeper of its Museum in 1826, his classic work of 1829 on the geology of the Yorkshire coast was both an advertisement for the Museum and its first fruit.[6] Though Phillips's career prospered nationally in the 1830s, York remained his base and Yorkshire his research area. He did publish on the lower coal measures, but the chief thrust of his work was the structure of the carboniferous limestone in the north west of the county.[7] It is significant that in 1837 he surveyed the Ingleton area of Yorkshire for coal as a commission and not on his own initiative; while surveying he was more interested in the Bowland anticlines than in locating coal.[8] Thus the two leading publishers on Yorkshire geology focused on the coast and the north-west dales and not on the coal-field.[9]

The non-mining geology pursued by Phillips and by the Yorkshire Philosophical Society was of course quite appropriate to an ancient cathedral city and county centre before the intrusion of Hudson's railways. Yet the philosophical societies of the major West Yorkshire manufacturing towns, all situated on the coal measures, followed this pattern of polite ornamental non-industrial geology. Of the active members of the Leeds Philosophical and Literary Society (f. 1818) only E. S. George had cultivated the geology of the local coal-field: when the Leeds Society launched its solitary volume of *Transactions* in 1837 the only paper on that subject was a posthumous one by George.[10] Yet it did show a brief spasm of interest in local coal geology in late 1837 just when the Geological and Polytechnic Society of the West Riding of Yorkshire was being formed. On 8 December 1837 J. G. Marshall capitalized on eight lectures on geology then being delivered in Leeds by James Finlay Weir Johnston, Reader in chemistry and mineralogy at the University of Durham, by establishing a geological group to cover Leeds geology and especially the coal-mining area of Middleton in south Leeds. After a brief flurry of activity and co-operation with T. W. Embleton, the Manager of the Middleton collieries, this Leeds geological group was defunct by 1839; and Marshall had already begun to study the metamorphic rocks of the Lake District where he had a large country house.[11] In Sheffield a similar lack of interest in coal-field geology was apparent. From its inception in 1822 to 1837, the Sheffield Literary and Philosophical Society mustered only Charles Morton, an engineer and colliery manager, with any sustained research interest in coal geology and mining technics.[12] By the late 1830s both the Sheffield and Leeds Societies were encouraging belles-lettres: local practical men were especially reluctant to give research level papers on local manufacturing and its technical problems, though without doubt they participated in the scientific enterprise as patrons, audience, and diffusers.[13] Wakefield was the most important town nearest to the centre of the worked coal-field, but there the question was the desperate one of survival: established in 1826 as a polite debating circle, the Wakefield Lit. and Phil. evaporated in 1838. Out of over 100 papers delivered to it, only three dealt with coal geology and mining problems.[14] At Bradford no regular scientific society existed until

1839 when William Sharp, senior surgeon at the Infirmary, established the first Bradford Philosophical Society.[15] It is therefore
indisputable that before 1837 the chief West Yorkshire Lit. and
Phils. in manufacturing towns actually on the county's large
coal-field gave scant institutional encouragement to research and
publication concerning its geology. This patent indifference to
geological knowledge concerning Yorkshire's most important
mineral resource was paralleled in the specialist geological societies
founded in the 1830s in the Irish and Scottish capitals (Dublin 1831;
Edinburgh 1834) which were devoted primarily to polite geology.
The former was founded and run chiefly by Dublin academics who
were not indifferent to Ireland's possible economic development
but were mainly interested in the geological structure of their
country. Likewise, the Edinburgh Society, composed of businessmen, existed 'to dignify and adorn their hours of recreation by
scientific pursuits'; accordingly it gave negligible attention to the
geology of the adjacent Lothian coal-field.[16]

There were, however, by 1837 two provincial specialist scientific
societies, the Royal Geological Society of Cornwall (f. 1814) and
the Natural History Society of Northumberland, Durham, and
Newcastle upon Tyne (f. 1829), which placed their main emphasis
on local geology with special reference to local mining. The former
published a mixture of pure geology, mineral analyses, mining
technology and statistics in its *Transactions*; and brought together
local gentry and mine owners to an extent unknown in the
Yorkshire literary and philosophical societies. Much of its success
was due to men such as John Henry Vivian, a cultivated gentleman
and mining entrepreneur who was familiar with the great continen-
tal mining academies. Even with this propitious social composition,
however, the Cornwall Society had not fulfilled its own stated aims
of completing a geological map of the county, of establishing a
mining records office, and of setting up a mining school. Retrospec-
tively the Cornish Society seems to have been as well placed as any
local geological society could have been to promote effective
co-operation between geologists, mine owners, and mine managers, and to encourage the study of local geology with particular
reference to mining problems. Its failure to meet its chief desiderata
shows that during the 1820s and 1830s such a programme faced
colossal difficulties.[17]

The Natural History Society of Northumberland nurtured similar
ambitions. Indeed, it was founded in 1829 as a splinter group from

the Newcastle Literary and Philosophical Society to give greater attention to the geology of the North-Eastern Coal-field, then the most productive in Britain. This Society's commitment to coal-field geology was shown particularly in its proposed map of the whole of Northumberland, Durham, and Cumberland, which was to be on such a scale that 'the out-crop of each principal bed of coal, sandstone, or limestone, shall be minutely laid down, together with the range and direction of the principal dykes and veins which intersect them, and this to be accompanied with various sections through the strata to the greatest depth ascertained by the several mines now in course of working'.[18] The chief protagonist of this ambitious programme was John Buddle, the mining engineer who had earned the sobriquet 'King of the coal trade' for his practice of dividing coal workings into panels and especially for his system of compound ventilation.[19] Buddle was worried that in different parts of the North-Eastern Coal-field, the same coal seams were called different names, so he pushed hard not only for a map showing dykes but also for sections showing the direction, bendings, and "throws" of these dykes. Buddle also was the prime mover of a scheme for making the Society's Museum a depot for the records of former workings of exhausted or relinquished collieries as a guide to posterity by which expense and accidents might be avoided.[20] Even though these two schemes were powerfully advocated by Buddle, neither was implemented in the 1830s by the Natural History Society of Northumberland. Sensing failure in Newcastle, Buddle turned to the British Association for the Advancement of Science when it visited Newcastle in 1838; having revealed his own map and sections of the Newcastle coal-field, he joined Thomas Sopwith in successfully urging the British government via the Association to establish the Mining Records Office which opened in London in 1840. As with the Cornwall Geological Society, the Newcastle one had a propitious social composition which embraced mining men led by Buddle, clerics such as William Turner, the aristocrat W. C. Trevelyan, the independent gentleman Selby, the land surveyor Hewitson, the lawyer Adamson, the insurance agent William Hutton, the medical bureaucrat Winch, and the provision merchant Alder.[21] Nevertheless most local mine owners and managers did not share Buddle's enthusiasm for making public the geological details of the Newcastle coal-field; hence the failure in the 1830s of the Newcastle Society, like that of the Cornwall one, to produce a local geological map and to establish a local mining records office.

Yet Buddle himself had shown that it was possible though difficult for one man to combine fruitfully economic geology and mining technics, while earning the plaudits of the metropolitan geological coterie: at the Newcastle meeting of the British Association for the Advancement of Science in 1838, he became one of the few mining men to hold a vice presidency in the geological section.

II

The programme and indeed title of what became the Geological and Polytechnic Society of the West Riding of Yorkshire were not easily defined. The Society was formed on 1 December 1837 with the naive Baconian inductivist aim of collecting, recording, and comparing geological and mechanical information pertinent to the Yorkshire coal measures and coal trade. Its geological model was the Natural History Society of Northumberland; its mechanical one was the Royal Polytechnic Society of Cornwall (f. 1833) which with the varied inducements of very low membership fee (5s.), of prizes, and of exhibitions, devoted itself to encouraging Cornish arts and industry, especially mining and pilchard fishing.[22] In order to give body to its general aims, the Yorkshire Society turned for advice to two leading Northern savants, John Phillips and James F. W. Johnston. The presence of Phillips at the Society's second meeting on 14 December 1837, and Johnston's absence from it, ensured that in the circular of 16 December accompanying the rules of the Society its title was 'The Geological Society of the West Riding of Yorkshire'; there was also more than a hint that Phillips was to be the great Baconian interpreter who would "methodize" the geological data to be accumulated by the Society. Johnston, however, gave different private advice: though himself interested in Yorkshire coal-field geology, he shrewdly sensed that given the intellectual resources of the Society the polytechnic programme might be less quickly exhausted than the geological one. He therefore recommended that the Society should devote itself to the practical working and economy of coal-mines, their products, and ancillary industries, as well as to the exclusively geological programme advocated by Phillips. Faced with these two different agenda, the Society stuck to its original aims: by spring 1838 the prospectus of the retitled 'Geological and Polytechnic Society of the West Riding of Yorkshire' made clear that the Society intended to cover more than specialist geology without embracing general science. Fortunately the pressure of Phillips's lecturing engagements prevented an embarrassing conflict with Johnston, who in a lecture in June 1838 set out the ideology of the Society with the entire concurrence and positive assistance of its leading members.[23]

The most striking feature of Johnston's lecture was its studied avoidance of polite gentlemanly geology: the very title which permitted him to embrace the triptych of geology, polytechnics, and philanthropy, was witness to his sustained directly utilitarian thrust. For the known part of the coal-field Johnston set out a very ambitious programme modelled on Buddle's work in the north east: the identity of Yorkshire coal seams, their disturbances, dislocations, and "heaves", plus the relations between the Yorkshire and Lancashire coal-fields, all to be recorded in maps and sections. For the unknown part of the coal-field, Johnston urged borings east of the known coal-field 'at the expense of a common fund' to discover seams at greater depths than hitherto worked. Given his own preoccupations, Johnston called for the compilation of statistics of coal-mining and for systematic chemical study of coal. Of course, he advocated a Museum not just for the preservation of fossils, but more as a repository for mining records past and present. In buttressing his desiderata with the example of Buddle, Johnston was clearly trying to conjure up a Yorkshire imitator from the assembled mining interest.

In expounding the polytechnic part of the Society's programme, Johnston pointed to the ventilation and draining of mines as key areas, and especially to the development of anemometers and the improvement of the Davy safety-lamp. He also advocated the related study of possible improvements in iron manufacture, starting with the geology of ironstone beds and continuing with the question of the hot blast which he strongly supported. Of course, Johnston again invoked Buddle, this time as a model mining engineer; but, rather surprisingly, he recommended as a model of experimental technical research and publication on the properties of iron, the work on the strength of materials for steam boilers done by Alexander Dallas Bache under the auspices of the Franklin Institute in Philadelphia. In contrast with the geological and polytechnic components of the Society's programme, the philanthropic one received scant attention from Johnston who recommended a paternalism designed to solve mine owners' problems with their labour forces, such as luddism and insubordination, allied to collecting local economic and medical statistics. His models here were John Taylor's views on the management of miners and the inquiries encouraged by the Statistical Society of London (f. 1834). The overall nature of this programme was clearly a far cry from the gentlemanly geology of Somerset House.[24] This difference was explicitly adumbrated by Thomas Wilson who told Lord Morpeth that the object of the YGS was 'not so much to cultivate *theoretical* geology, as to investigate thoroughly the mineral seams, both of coal and ironstone, that cover so large a portion of this Riding, with reference to their *economic* value'.[25] The programme of the YGS also implied that it did not intend to compete with the general scientific societies in the area.

The Society's most successful effort to implement its ambitious aims was its published *Proceedings*. Here the YGS scored a coup. Clearly it could not rival the Geological Society of London which published both *Transactions* and *Proceedings*; but it did surpass the West Yorkshire Lit. and Phils. who either published memoirs sparsely or not at all. Its *Proceedings* appeared regularly from late 1839, showing a commitment to the advancement of science which the local general scientific societies could not equal. From its inception to 1843, the Society managed to mount its statutory four meetings per year, before declining to three a year in 1844, and to two a year in 1847. The papers published or delivered between 1838 and 1843 represent therefore the high point of the Society's activity. To what extent, however, did they implement the Society's programme?

Of the seventy-three contributions made in six years, a third was devoted to mining technics, including ventilation, safety-lamps, wire ropes, boilers of steam engines, and mining waste, the leading contributors being mainly West Yorkshire coal-mine or iron-furnace owners and managers, with the clear exception of Sopwith, the Newcastle land and mining surveyor, who recommended his pet subjects of isometrical projection and models of mining districts. The second most popular category was the geology of the West Yorkshire coal-field to which twenty-two papers were devoted, the main contributors again being mine and furnace men who clearly took seriously Johnston's desiderata for the known part of the coal-field. Problems of iron manufacture, mainly dealing with the relative merits of iron produced by the hot and cold air blast furnaces, attracted eight papers almost entirely given by local iron-masters. Agriculture, especially in its putative connections with geology, was energetically promoted by Thorp. In general, about two-thirds of all papers dealt with the geology of the coal-field and the technics of the coal industry. Yet already polite geology had made a small appearance: J. T. Clay had been quick to apply Agassiz's glaciation theory, promulgated at the Glasgow meeting of the British Association in 1840, to the questions of Yorkshire erratic boulders and "drift" gravel; and Sopwith expounded the glacial theory warmly. Even so, this intrusion of polite geology, inspired by Agassiz, hardly disturbed the dominance of coal-field papers, especially as general science other than geology and chemistry was totally unrepresented.[26]

In two other respects, however, the Society was less successful in achieving its manifest purposes during its opening six years. Though guided by Phillips and especially by Greenough, and having good local exemplars at Leeds and Newcastle to copy, the Museum of the Society, which cut across the local museums, did not prosper as much as its leaders wished. Earl Fitzwilliam, the President, had launched the Museum with a generous donation of money and specimens; but by 1841 Embleton, the Honorary Curator, in deploring the lack of donors of specimens, set out a list of desiderata to be fulfilled. To implement these aims, Martin Simpson, an expert on Yorkshire coast fossils, was appointed Curator at £50 per annum from summer 1842 to summer 1843 when the Society decided to put its Museum (cases of specimens in temporary accommodation in Wakefield) into the care of a West Yorkshire Lit. and Phil. because it could not afford a salaried Curator. By summer 1844 the YGS Museum had effectively been combined with that at Leeds. While many Lit. and Phils. existed primarily to support their museums and not to publish papers, the reverse was the case with the YGS.[27]

The second instance of incomplete success concerned the collaborative research by members of the Society, in co-operation with the Manchester Geological Society (f. 1838), on a section across the Pennine chain which would show the mutual relations of the Yorkshire and Lancashire coal-fields.[28] Greenough, an old-style mapper, then completing yet another geological map of England and Wales, supported the project warmly, urged exclusive concentration on the coal-field part of Yorkshire, and recommended a common vertical but not horizontal scale for the section.[29] After considerable discussion involving chiefly Morton, Embleton, Hartop and Thorp, by summer 1841 the line and scales of the section were decided with the revised aim of procuring that section most illustrative of the Yorkshire coal-field rather than that permitting the best comparison between the two coal-fields. The section across the Yorkshire coal-field was almost complete by late 1843, being apparently available for reference but not published.[30] As had happened at Newcastle with Buddle, it was left to an interested individual, Thorp, to publish *c.* 1847 a series of sections of the Yorkshire coal-field in connection with a projected work on it. The Society thus did not entirely succeed in producing a map and sections of the coal-field; and, concomitantly, it did not establish a depository of mining records.[31]

III

Even so, it is clear that between 1838 and 1843 the YGS, of which ordinary membership was restricted to West Riding residents, avoided the precariousness which afflicted much of the Lit. and Phil. movement from about 1840. This was chiefly due to a core-group of members: Thomas Wilson, the Secretary and Treasurer until 1842, who administrated the Society with high-minded zeal and relentless efficiency; and a small group of performers, that is men who gave papers at the quarterly meetings and promoted the Society's collaborative research. By late 1843 the Society nominally totalled 299 members, yet a quintet of performers (Thorp, Hartop, Embleton, Morton, Briggs) conducted by Wilson, constituted the small band largely responsible for the Society's fortunes.[32] Their most signal characteristic was their involvement in the Yorkshire coal and iron trade, chiefly the former, as mine owners or managers mainly in the Wakefield–Barnsley area; the Reverend Thorp, it should be noted, being a partner in a colliery as well as vicar of Womersley. Wilson, Briggs, Morton and probably Embleton, were mainstays of the West Yorkshire Coal Owners Association. Indeed, it was at a meeting of this Association at Wakefield on 1 December 1837 that the YGS was established. Unfortunately very little is known about the Association which was apparently devoted to problems of negotiating with colliers and to the maintenance of coal prices, at a time when the Yorkshire coal trade was exposed to great competition. Though its title proclaimed inclusiveness, only about one fifth of all Yorkshire coal owners were members of it.[33] There is unfortunately no evidence that the Association *per se* tried to map the county's coal measures, to establish a depot for local mining records, or to promote investigations into technical mining problems such as ventilation. Politically it was dominated by ardent free-traders and Liberals who, as defenders of commercial and civil liberty, vehemently opposed the state interference and espionage endorsed in the Mines Act of 1842.[34]

The YGS, as an offshoot of a trade association, was initially controlled by coal and iron men; not by a mixture of academics, independent gentlemen, medics, clerics, lawyers and merchants, as in other geological and scientific societies. Some of them had indeed a strong economic motive. In the 1830s Wilson, who had previously worked shallow seams at Silkstone, staked his capital in trying to find coal north-eastwards at Darton at greater depths than then usually contemplated; this risky venture depended on assumptions he made about the identity and direction of certain coal seams. Unfortunately his Kexborough pit, near Darton, failed disastrously because the shaft was sunk just where a "throw" occurred, leaving him no alternative but to withdraw from mining and from the secretaryship of the YGS in 1842.[35] His close friends Briggs and Morton were, however, successful in sinking pits eastwards. In 1836 Briggs sank the first ever successful pit through the magnesian limestone to the coal underneath at Newton near Castleford. From 1841 in partnership with Morton he successfully worked the Whitwood collieries near Normanton, which were deeper than his Flockton pits and adjacent to the new Sheffield–Leeds railway. Embleton, too, was obsessed with the question of identity of coal seams not least because the near exhausted seams of the Middleton colliery near Leeds, which he managed from 1830, prevented him from competing with the more successful Barnsley pits.[36] There is, then, enough evidence to suggest that the majority of these mine owners and managers, at least half of whom were working deeper mines than average, hoped that, if a comprehensive summary in some form of the relations of the various seams could be compiled from previously scattered local information, then the sinking of new deeper shafts would be less hazardous financially. No doubt the bigger capitalists among them hoped to acquire knowledge of the deeper seams which only they had the financial capacity to exploit. No doubt, too, the financial stringency, falling prices, unemployment, and labour troubles which erupted in 1837 combined to strengthen their awareness of the financial risks they faced.[37]

Given this vital economic motive, it is not surprising that most of the mining core-group enjoyed few contacts with organized polite science in either the metropolis or in the provinces. None was FRS,

FGS, or connected with the Yorkshire Philosophical Society. With the single exception of Morton who worked conspicuously in the Sheffield Literary and Philosophical Society from 1833 to 1838, they were not active in their local Lit. and Phil. From 1839 the YGS was dependent on these local scientific societies because it used their premises as it moved from town to town for its meetings; yet Wilson believed that most provincial scientific societies were 'powerless to advance the bounds of science, or even to communicate what is already known . . .'.[38] With the solitary exception of Hartop, none was known before 1837 by his publications or by appearances at public gatherings of scientists such as the meetings of the British Association. Hartop had given papers to the Association on the geology of the Don valley which resulted in an unimplemented resolution about the desirability of sections and plans of the South-West Yorkshire coal-field.[39] Their relative isolation before 1837 was not, however, the result of educational inferiority. After all, Wilson and Thorp were Cambridge graduates, while Morton and perhaps Embleton had spent a year at Edinburgh University, the others being probably autodidactic and trained by apprenticeship. In short, they formed a distinct and coherent group because of their occupational interest in mining geology and its associated technical problems such as ventilation and ropes.

In the opening phase of 1837–43, the managers of the YGS contrived to induce local West Yorkshire residents to be the Society's performers. This was largely due to the incessant zeal shown by Wilson in soliciting papers. In May 1841, for instance, the prospects for the June Leeds meeting looked gloomy: 'I have written to ask Mr. Teale, who cannot help; Mr. West, who does not answer; Mr. Holt, who says he has not data. . . . Could Mr. Chantrell be spurred up; Mr. Thorp says flatly he has not time . . . . Our brethren at Manchester are I think ahead of us now, they seem to have no difficulty in getting papers.'[40] With the exception of the core-group, mine and furnace managers and engineers certainly did not stampede to offer papers. Rather they had to be cajoled by Wilson who publicly regretted that most members, feeling that only a long and elaborate paper was appropriate for a learned society, were deterred from making brief factual communications.[41]

At the same time the YGS leaders were concerned to create, establish, and if possible expand the local audience for its performers. This was a considerable problem because the membership was scattered over the West Riding and subscriptions were difficult to collect. Furthermore, while lecture courses were by then the staple diet of Lit. and Phils. the YGS disdained this stratagem.[42] It tried, however, to increase its membership of non-performers by offering a low subscription rate of half a guinea, and from 1839 by holding its meetings in turn in all the major towns of the Riding. Given the long-standing interest shown by the gentry in science if it was orientated towards agriculture, it was inevitable that the YGS should try from 1840 to attract the landed interest by cultivating agricultural geology and agricultural chemistry, and by running joint meetings with the Yorkshire Agricultural Society (f. 1837). Wilson himself had grave doubts about the utility of geology in reference to agriculture, yet in 1841 he even contemplated a merger with the Yorkshire Agricultural Society because the popularity of agriculture would ensure a supply of funds sufficient to implement the YGS's aims and to procure paid research staff.[43] A further device Wilson wished to employ was to induce non-local geological stars to attend meetings, or even to perform. He kept looking to Phillips as a visitor from York who would bolster both attendance and meetings. Embleton, however, viewed the idea of importing star performers as ultimately debilitating. As he told Wilson, 'Prof Phillips' absence [from the Leeds meeting of December 1839] will be the means of lessening the attendance at the evening meeting, but I think we ought always to depend upon papers from the members instead of relying upon the assistance of "Professors of Geology".'[44] In the opening six years, Embleton's view prevailed: Johnston, the expositor of the Society's aims, and Sopwith, a mining engineer, were the solitary imported star performers, both from Newcastle; while metropolitan and Oxbridge lions, such as Sedgwick, Buckland and Greenough, accepted invitations to "roar" merely as discussants at meetings, of which the annual ones were arranged to fall near the times of meeting of the British Association in order to capture its stars. Wilson caught such geological giants chiefly through his friend Sopwith and the President Earl Fitzwilliam whose scientific tastes, menagerie, pits, furnaces, and lavish hospitality made his house at Wentworth near Rotherham a favourite port-of-call for passing geologists. Indeed, the YGS did not scruple to put the date of its autumn 1840 meeting into Sedgwick's hands in order to ensure his presence.[45]

For advice on its Museum the YGS turned outside the Riding to Phillips and to Greenough; and for guidance on its proposed section of the Yorkshire coal-field primarily to Greenough and secondarily to Sopwith. Clearly the YGS felt the necessity of taking outside advice only from selected individuals who had a particularly appropriate experience, such as Greenough's in mapping, Sopwith's in section work, and Phillips's in curatorship, which could further the Society's aim of cultivating not theoretical but economic geology. With respect to institutions as well as individuals outside the Riding, the YGS cultivated sturdy northern self-help as much as possible. It collaborated with the Manchester Geological Society on the projected Pennine section work; and perforce it occasionally joined forces with the Yorkshire Agricultural Society with which none the less it refused to be amalgamated. While it never contemplated imitating the Geological Society of London, with which its contacts were negligible, it did support the British Association with whose geological and statistical sections it shared some common aims.

IV

The second phase of the Society's history covered the years 1844–6 inclusive, when only three out of the statutory four meetings per year were held. This lapse was caused partly by two changes of Secretary: in 1842 Clay replaced Wilson, but owing to family bereavement and growing business responsibilities he resigned in 1844. Morton and William West were apparently approached to succeed Clay but both refused. Thorp came to the rescue, being assisted from 1845 by Henry Denny, Curator of the Museum of the Leeds Phil. and Lit. and Lecturer in botany in the Leeds school of medicine.[46] In any event, these three years witnessed distinct changes in the areas covered by papers and in the identity of performers, as well as a decline in productivity from about twelve to nine papers a year.

The most obvious shift was the growth of general science and of general geology, with the associated decline of mining technics, of coal-field geology, and of iron manufacture. Indeed, a third of the papers was devoted to general science, presumably to avoid lacunae caused by the relative dearth of papers on mining geology and technics. By far the leading general contributor was William Scoresby, the vicar of Bradford from 1839 to 1847, who could always be relied upon for interesting experimental demonstrations concerning magnetism and indeed for extemporaneous communications. He was supported by William Sykes Ward, John

Deakin Heaton, and William West, three Leeds men who were all key members of the Leeds Philosophical and Literary Society. Through Scoresby and the Leeds trio, polite general science, which was an anathema to the early managers, became the leading area covered. Mining technics, especially ventilation, were represented on a smaller scale, the leading performers of the previous phase remaining silent. There was a similar decline in Yorkshire coal-field geology, Embleton, Morton and Briggs offering no papers but contributing to discussion. Only one third of the papers dealt with mining geology and technics, the two chief areas purportedly cultivated by the Society; and only Thorp of the first core-group continued to perform regularly. Hartop, Embleton, Morton and Briggs, who gave twenty-three papers in the opening six years, produced only two contributions in the subsequent three years. While the coal men were becoming inactive, Yorkshire non-coal geology and general geology were promoted by two imports, Phillips and Edward Charlesworth, who were successive curators of the York Museum. Clearly in this second phase, the coal and iron men who had previously run the Society had lost intellectual steam or enthusiasm, or developed new interests and careers. As the Society imported few star performers from outside the Riding – York excepted – inevitably it had to transform itself partly into a peripatetic Lit. and Phil. offering an eclectic menu of polite general science and ornamental geology as well as mining geology and technics. Hence it depended increasingly on such local performers in the Yorkshire Lit. and Phil. and Mechanics' Institute networks as Scoresby, Phillips, West and Ward, none of which was occupationally involved in the coal trade.[47]

The third phase of the Society's fortunes covered 1847–53 when no more than two meetings per year were staged. Indeed, by 1854 it became almost defunct.[48] This was partly due to the migration in 1848 of Thorp, the Secretary and Treasurer, from Womersley near Pontefract to the vicarage of Misson in North Nottinghamshire, where he was inconveniently distant from the Society's area of operation. Once again there was a distinct shift in the areas covered by papers and in the identity of performers, as well as a further decline in productivity to about six papers per year. The most striking feature was the almost total disappearance of coal-field geology, the previous stalwarts being entirely inactive. Yorkshire non-coal geology and general geology flourished, being dominated by the wealthy Sheffield devotee Henry Clifton Sorby who favoured the Society with important pioneering papers on petrology.[49] Mining technics were, however, the most popular field, ventilation being the key problem covered: Thorp, the former doyen of coal-field geology, gave more attention in these years to mining technics. A much increased interest in applied science was largely due to Ward, who also revived interest in iron manufacturing. General science lost its previous dominance, becoming the exclusive domain of Leeds savants. What happened *vis à vis* performers was that Scoresby and Phillips, having left Yorkshire, were not available; only Thorp of the original core mine-owning group survived actively into the third phase. Not surprisingly the Society became even more dependent on Leeds men such as Ward, West, Denny and Thomas John Pearsall, and on such Sheffield savants as Sorby and James Haywood.[50] With the exception of Pearsall, all these men were active as performers and as administrators in their two respective Lit. and Phils. There was indeed some overlap of papers; but the YGS was saved from being a mere extension of the Leeds and Sheffield Societies because unlike them it published *Proceedings* and gladly offered an outlet for papers on applied science. None of this third core-group, with the exception of Thorp, was engaged in the coal trade. The utter collapse of the coal-field research programme was confirmed by the total absence of any paper on this topic between 1850 and 1853 inclusive; and it is significant that the polytechnic programme was maintained not only through the steady support of mining men such as Thorp and Benjamin Biram, but also through that given by non-mining savants such as Ward and West.[51] In short, of Johnston's three chief desiderata adumbrated on behalf of the Society, the polytechnic programme was partly implemented, though not with exclusive regard to the coal and iron trades; after a vigorous start, the coal-field geology programme entirely collapsed; while the philanthropic one was never even launched properly. At the same time, as the mining interest withdrew from giving papers to the Society, it became more and more dependent on local savants in the two major West Yorkshire towns. By June 1854 even Thorp thought the Society should wind up because of few contributors, slack members, poor attendance, and debt through having to pay rent to the Leeds Phil. and Lit. as well as a salary to Denny as Assistant Secretary. Shortly afterwards his resignation as Secretary and replacement by Ward signalled the extinction of the mining interest in running the Society.

V

For all of the Society's more active performers, the *Proceedings* was their sole or chief publishing outlet for their papers on geology, at a time when the Leeds Phil. and Lit. published no papers and the Sheffield one offered abstracts of papers only from 1850. This was, one suspects, a strong inducement to savants in those two towns, such as Ward, West and Sorby, to support the YGS as well as their local scientific society. Yet the demise of the mining geology programme after such a vigorous start calls for explanation. There was, of course, the sheer difficulty of the geological problems involved, basically due to the bending of the coal seams, their varying thickness, and the faults which dislocate them, to which attention was often drawn. On such an important matter as the geology of the Don valley the Society produced no consensus. There was also the vexing problem of inducing mine owners in general to proffer detailed local information. The Society's frequent calls for mining facts were widely ignored, partly because of the pervasive belief in the coal trade that secrecy served competitive capitalism best: Wilson regarded the proposed government mining inspectorate as a system of espionage; and probably many local mine owners regarded the Society's call for the production of accurate and detailed local maps, plans, sections, models and records, as another form of espionage. For these men in particular, economic entrepreneurship was incompatible with Baconian co-operative inductivism. In any event, many mines possessed no written records of important features such as the amount and direction of the throw of a fault; and when sections existed, they were sometimes merely general.[52] Though many mine owners did join the Society, their membership was mainly passive and nominal: only a small proportion of the mining membership was prepared to divulge and able to discuss local geological information. The active core of mining men who were performers was unable to renew itself by recruiting converts from the coal trade, and was consequently small. Why did these men, initially prolific, fail to sustain their programme?

It seems that for men of merely local knowledge the pursuit of local geology, like extractive mining, was subject to the law of diminishing returns; practical men such as Embleton, Briggs and Hartop apparently had geological knowledge of only a particular part of the coal-field, leaving only Thorp and probably Morton who possessed more than such local information. Hence most of the active mining men simply ran out of data when they had described their own immediate locality. Perhaps the vehicle of a voluntary learned society was inappropriate for a programme which was concerned with ambitious industrial research and development: certainly some of the active mining coterie diverted their attention to other fields, Briggs developing an interest in agriculture and Morton a career as a mines inspector; Embleton, who moved into mining consultancy, probably became disenchanted with the Society and its Museum; and in 1844 Hartop left Sheffield and the baffling study of the river Don's geology to manage the Bowling iron works at Bradford. One also surmises that Wilson's spectacular financial failure as a mine owner did not convince the practical men who had warned him against disaster that it was in their interest to join his Society. On the other hand, Briggs's pioneering ventures in deeper mining were economically successful without apparently drawing on the geological work done by the YGS. For all these reasons the Society could not sustain a programme devoted to creating a body of geological knowledge applicable to coal-mining, though its polytechnic work revealed the greater feasibility of applying scientific procedures to the engineering problems of the industry.[53] In sum, this most explicit utilitarian and economically motivated Society, active in an industrial and mining area, succumbed considerably as a *pis aller* to the lure of science as ornamental learning.

Table 4   *Topics and number of papers presented to the YGS, 1838–53*

Topics	1838–43		1844–6		1847–53	
	no. of papers	papers /yr	papers	papers /yr	papers	papers /yr
West Yorkshire coal geology	22	3.7	4	1.3	1	0.1
West Yorkshire non-coal geology	4	0.7	1	0.3	6	0.9
General geology	2	0.3	4	1.3	3	0.4
Mining technics	24	4.0	5	1.7	10	1.4
Iron manufacture	8	1.3	0	0.0	3	0.4
General technics	0	0.0	1	0.3	8	1.1
Coal chemistry '	2	0.3	1	0.3	1	0.1
Non-coal chemistry	1	0.2	0	0.0	2	0.3
Agriculture	6	1.0	2	0.7	4	0.6
Philanthropy, statistics and public health	1	0.2	1	0.3	2	0.3
General science	0	0.0	8	2.7	4	0.6
Architecture	3	0.5	1	0.3	0	0.0
Totals	73	12.2	28	9.3	44	6.3

Table 5   *Topics and authors of papers presented to the YGS, 1838–53*

Topics	1838–43 Author (no. of papers)	1844–6 Author (no. of papers)	1847–53 Author (no. of papers)
West Yorkshire coal geology	Thorp (10) Embleton (4) Morton (3) Hartop (2) Briggs (1) Simpson (1) Teale (1)	Thorp (2) Hartop (1) Denny (1)	Denny (1)
West Yorkshire non-coal geology	Clay (2) Alexander (1) Lee (1)	Phillips (1)	Sorby (3) Denny (1) Thorp (1) T. West (1)
General geology	Sopwith (1) Mackintosh (1)	Binney (1) Charlesworth (1) Phillips (1) Solly (1)	Sorby (3)

Table 5 – *continued*

Topics	1838–43 Author (no. of papers)	1844–6 Author (no. of papers)	1847–53 Author (no. of papers)
Mining technics	Hartop (4) Morton (3) Biram (2) Briggs (2) Embleton (2) Fourness (2) Sopwith (2) Fletcher (1) Hanson (1) Holt (1) Holmes (1) Lucas (1) Roberts (1) Ward (1)	Barker (1) Biram (1) Fourness (1) Ward (1) West (1)	Thorp (3) Nasymth (2) Biram (1) Bodington (1) Ramsden (1) Ward (1) West (1)
Iron Manufacture	Hartop (2) Todd (2) Graham (1) Leah (1) Scoresby (1) Solly (1)	——————	Ward (2) Solly (1)
General technics	——————	Roberts (1)	Ward (4) Broadrick (1) Dalton (1) Dresser (1) Pearsall (1)
Coal chemistry	West (2)	Lucas (1)	W. L. Simpson (1)
Non-coal Chemistry	West (1)	——————	Haywood (1) West (1)
Agriculture	Thorp (5) Hamerton (1)	Briggs (1) Haywood (1)	Briggs (1) Haywood (1) Thorp (1) Wilkinson (1)
Philanthropy	Nowell (1)	Thorp (1)	Alexander (1) Haywood (1)
General science	——————	Scoresby (5) Heaton (1) Ward (1) West (1)	Ward (2) Heaton (1) Pearsall (1)
Architecture	Wallen (3)	Wallen (1)	——————

Acknowledgements

For valuable help and criticism I am indebted to Stella Butler, Ian Inkster, John Pickstone, Roy Porter, and to the members of seminars at the Universities of Leeds, Oxford, Pennsylvania, and Johns Hopkins, to whom a preliminary version was given. For permission to cite unpublished sources, I am grateful to the Council of the Yorkshire Geological Society (archives, minute books), the University Library Cambridge (Greenough papers), Leeds Public Libraries Archives (Wilson papers), the Yorkshire Philosophical Society (Council minutes), and the Leeds Philosophical and Literary Society (minute book of transactions).

Notes and references

(Details of individuals are given only for those not recorded in *DNB*, *DAB*, or *DSB*.)

1 A. W. Thackray, 'Natural knowledge in cultural context: the Manchester model', *American Historical Review*, 1974, **79**, 672–709; S. Shapin, 'The Pottery Philosophical Society, 1819–1835: an examination of the cultural uses of provincial science', *Science Studies*, 1972, **2**, 311–36.

2 R. S. Porter, 'The industrial revolution and the rise of the science of geology', in M. Teich and R. M. Young (eds.), *Changing Perspectives in the History of Science*, London, 1973, pp. 320–43.

3 J. W. Davis, *History of the Yorkshire Geological and Polytechnic Society, 1837–1887. With Biographical Notices of Some of its Members*, Halifax, 1889 (henceforth *Davis*). Yorkshire was then divided into three parts called Ridings.

4 J. B. Morrell, 'London institutions and Lyell's career: 1820–41', *British Journal for History of Science*, 1976, **9**, 132–46; R. Burt, *John Taylor: Mining Entrepreneur and Engineer 1779–1863*, Buxton, 1977; R. S. Porter, 'Gentlemen and geology: the emergence of a scientific career, 1660–1920', *The Historical Journal*, 1978, **21**, 809–36.

5 W. Buckland, *Reliquiae Diluvianae*, London, 1823, pp. 1–51; J. W. Clark and T. M. Hughes, *The Life and Letters of the Reverend Adam Sedgwick*, Cambridge, 1890, **i**, pp. 294–7, 531, and **ii**, p. 503; T. Sheppard, *Bibliography of Yorkshire Geology*, London, Hull and York, 1915, which is *Proceedings of the Yorkshire Geological Society*, 1915, **18** (henceforth *PYGS*).

6 A. D. Orange, *Philosophers and Provincials: The Yorkshire Philo-

sophical Society from 1822 to 1844, York, 1973; J. Phillips, *Illustrations of the Geology of Yorkshire*: part 1, *The Yorkshire Coast*, York, 1829.

7   cf. Phillips 'On the lower or ganister coal series of Yorkshire', *Philosophical Magazine*, 1832, **1**, 349–53; and *Illustrations of the Geology of Yorkshire*: part 2, *The Mountain Limestone District*, London, 1836.

8   J. Phillips, *A Report on the Probability of the Occurrence of Coal and Other Minerals in the Vicinity of Lancaster. Addressed to the Lancaster Mining Company*, Lancaster, 1837.

9   H. C. Versey, 'History of Yorkshire geology', *PYGS*, 1973–6, **40**, 335–52 (338).

10   E. S. George, 'On the Yorkshire coal-field', *Transactions of the Leeds Philosophical and Literary Society*, 1837, **1**, 135–91; published annual *Reports of the Council of the Leeds Philosophical and Literary Society*, Leeds, 1820–38. On Edward Sanderson George (1801–30), Leeds chemical manufacturer, see E. K. Clark, *The history of . . . the Leeds Philosophical and Literary Society*, Leeds, 1924, pp. 12–13, 29, 42, 146.

11   General minute book of transactions of the Leeds Philosophical and Literary Society, 1821–41 (entry for 8 December 1837); *Eighteenth Report of Council of Leeds PLS*, Leeds, 1838, pp. 8–10; *Nineteenth Report of Council of Leeds PLS*, Leeds, 1839, p. 11; *Twentieth Report of Council of Leeds PLS*, Leeds, 1840, p. 6; *Davis*, pp. 180–2; W. G. Rimmer, *Marshalls of Leeds, Flax-Spinners 1788–1886*, Cambridge, 1960, pp. 182, 221–2; *The Athenaeum*, 1839, p. 646. James Garth Marshall (1802–73); Thomas William Embleton (1809–93), mine manager, obituary in *PYGS*, 1892–4, **12**, 335–9.

12   Published *Annual Reports of the Sheffield Literary and Philosophical Society*, Sheffield, 1824–38; published *Annual Reports of the Sheffield Mechanics' Institute*, Sheffield, 1833–7. On Morton (1811–82), see *Davis*, pp. 61–3; J. Goodchild, 'The first mines inspector in Yorkshire. Part I', *South Yorkshire Journal of Economics and Social History*, 1971, part 3, 15–17.

13   *Seventeenth Report of the Sheffield LPS*, Sheffield, 1840, p. 4; I. Inkster, 'The development of a scientific community in Sheffield, 1790–1850: a network of people and interests', *Transactions of the Hunter Archaeological Society*, 1973, **10**, 99–131.

14   Programmes for 1826–36 held in Wakefield City Archives; *West Riding Herald*, 13 January 1837 and 28 April 1837.

15   J. James, *The History of Bradford and Its Parish*, London and

Bradford, 1866, pp. 245–8; W. Scruton, *Pen and Pencil Pictures of Old Bradford*, Bradford, 1889, pp. 90–3.

16   G. L. Davies, 'The Geological Society of Dublin and the Royal Geological Society of Ireland 1831–1890', *Hermathena*, 1965, no. c, 66–76; 'Memoir of the Society', *Transactions of the Edinburgh Geological Society*, 1870, **1**, 1–6 (1).

17   R. S. Porter, *The Making of Geology: Earth Science in Britain, 1660–1815*, Cambridge, 1977, pp. 134–5. John Henry Vivian (1785–1855) was head of Vivians, the copper smelters.

18   T. R. Goddard, *History of the Natural History Society of Northumberland, Durham and Newcastle upon Tyne 1829–1929*, Newcastle, 1929, p. 39.

19   See the copious evidence of Buddle printed in *Report from the Select Committee on Accidents in Mines*, Parliamentary Papers, 1835, **5**, 1–373 (131–88, 215–23).

20   J. Buddle, 'Synopsis of the several seams of coal in the Newcastle district', *Transactions of the Natural History Society of Northumberland, Durham and Newcastle upon Tyne*, 1831, **1**, 215–24; *idem*, 'Reference to the sections of the strata of the Newcastle coal field', ibid., 225–40; *idem*, 'Suggestions for making the Natural History Society a place of deposit for the mining records of the district', paper read at a meeting of the Natural History Society of Northumberland, Durham, and Newcastle upon Tyne, 23 December 1834, and printed in their *Transactions*, Newcastle, 1838.

21   Buddle, 'Observations on the Newcastle coal-field', *Report of the Eighth Meeting of the British Association for the Advancement of Science*, London, 1839, pp. 74–6; ibid., p. xxiii; *Report of the Ninth Meeting of the British Association for the Advancement of Science*, London, 1840, p. 174.

22   *Second Annual Report of the Cornwall Polytechnic Society*, Falmouth, 1834, pp. 9–13; G. W. Roderick and M. D. Stephens, *Science and Technical Education in Nineteenth-Century England*, Newton Abbot, 1972, pp. 119–33; *Davis*, pp. 2–4.

23   *Davis*, pp. 3–14; Wilson to Phillips, 9 December 1837; Wilson to Johnston, 22 January 1838; Wilson to Phillips, 10 February 1838; Wilson to Johnston, 5 March 1838, in Wilson Papers, DB 178/23.

24   J. F. W. Johnston, *The Economy of a Coal-Field: An Exposition of the Objects of the Geological and Polytechnic Society of the West Riding of Yorkshire, and of the best means of attaining them*, Durham, 1838. For Bache's work see B. Sinclair, *Philadelphia's*

Philosopher Mechanics: A History of the Franklin Institute 1824–1865, Baltimore, 1974.

25   10 July 1839, Wilson papers, DB 178/23, italics in original. Thomas Wilson (1800–76), *Davis*, pp. 49–52.

26   Tables 4 and 5, pp. 249 and 250. Clay, 'Observations on the occurence of boulders of granite, and other crystalline rocks, in the valley of the Calder, near Halifax', *PYGS*, 1839–42, **1**, 201–6 (read June 1841), and 'Observations on the Yorkshire drift and gravel', ibid., 338–51 (read December 1841); Sopwith, 'On the evidence of the former existence of glaciers in Great Britain', ibid., 419–42 (read March 1842). Joseph Travis Clay (1805–92) was a Quaker who founded the worsted coat trade in Brighouse.

27   *Davis*, pp. 150–67; J. E. Hemingway, 'Martin Simpson, geologist and curator', in H. B. Browne, *Chapters of Whitby History 1823–1946*, Hull and London, 1946, pp. 93–105.

28   For the early Manchester Society see R. H. Kargon, *Science in Victorian Manchester: Enterprise and Expertise*, Baltimore, 1977, pp. 24–7, 30–4.

29   Wilson to Greenough, 26 October 1839, Greenough Papers, Add 7918(8); Greenough to Wilson, 4 November 1839, Wilson papers, DB 178/27.

30   *Davis*, pp. 89–118. On Henry Hartop (1786–1865), furnace manager, see G. Mee, *Aristocratic enterprise: The Fitzwilliam Industrial Undertakings 1795–1857*, Glasgow, 1975, pp. 45–63; Reverend William Thorp (1804–60), *Davis*, pp. 213–15, 382–6.

31   YGS Council minutes, 18 December 1839. Thorp's map survives in Wakefield City Archives.

32   Henry Briggs (1797–1868) for details of whom I am indebted to an unpublished typescript by Mr J. Goodchild, Curator of Wakefield City Archives; *Davis*, pp. 60–1.

33   Evidence of Wilson in *Children's Employment Commission*: *Appendix to First Report of Commissioners. Mines. Part I.* Parliamentary Papers, 1842, **16**, 205.

34   Wilson to Committee of Yorkshire coal owners, 7 February 1842, Wilson Papers, DB 178/22; circular of Wilson to coal owners of the West Riding of Yorkshire, 26 May 1842, Wilson Papers, DB 178/27; *Report of the Committee Appointed at a Meeting of the Yorkshire Coal-Owners...on 21 May 1841, to Take into Consideration the Commission of Enquiry into the Employment of Children and Young Persons in Mines and Manufactories...*, Barnsley, June 1841, Wilson Papers, DB 178/27.

35 Account of Wilson in unpublished volume 2 of J. Wilkinson, 'Worthies, families and charities of Barnsley and the district', London, n.d., in Wakefield City Archives.

36 W. G. Rimmer, 'Middleton colliery, near Leeds, 1770–1830', *Yorkshire Bulletin of Economic and Social Research*, 1955, **7**, 41–58.

37 A. D. Gayer, W. W. Rostow and A. J. Schwartz, *The Growth and Fluctuation of the British Economy 1790–1850*, Oxford, 1953, **i**, 242–76.

38 Wilson to secretaries of Yorkshire Philosophical Society, 9 February 1842, Wilson papers DB 178/23, reproduced in YPS Council minutes, 14 February 1842.

39 *1833 Report of the British Association*, xxxiv, 478; *Proceedings of the Fifth Meeting of the British Association for the Advancement of Science, held in Dublin*, Dublin, 1835, 107.

40 *Davis*, pp. 172–3. Thomas Pridgin Teale (1800–67), Leeds surgeon, *Davis*, pp. 54–8; William West (1792–1851), Leeds chemical consultant, *Davis*, pp. 238–40; Henry Holt (1812–69), Wakefield mining engineer and surveyor, *Davis*, pp. 52–3; Robert Dennis Chantrell (1793–1872), Leeds architect.

41 *PYGS*, 1839–42, **1**, 37.

42 Wilson to Murray, 5 October 1838, Wilson Papers, DB 178/23.

43 Wilson to Fitzwilliam, 31 December 1840; Wilson to Johnston, 3 January 1841; Wilson to Fitzwilliam, 21 January 1841, all in Wilson Papers, DB 178/22; Wilson to Fitzwilliam, 9 May 1841, Wilson Papers, DB 178/28.

44 YGS Council minutes, 1 November 1839; Embleton to Wilson, 13 November 1839, YGS Archives.

45 *PYGS*, 1839–42, **1**, 79–81, 338; ibid., 1842–8, **2**, 41; *Davis*, pp. 81, 115, 171–3, 226; YGS Council minutes, 8 August 1838, 8 February 1839.

46 YGS Council minutes, 28 May 1844.

47 Tables 4 and 5, pp. 249 and 250. William Sykes Ward (1813–85), Leeds lawyer and inventor, *Davis*, pp. 241–2; John Deakin Heaton (1817–80), Leeds physician, T. W. Reid, *A Memoir of John Deakin Heaton*, London, 1883; Edward Charlesworth (1813–93); T. Stamp and C. Stamp, *William Scoresby: Arctic Scientist*, Whitby, 1976.

48 *Davis*, pp. 259–60; Thorp to Wilson, 19 June 1854, Wilson Papers, DB 178/30.

49 Tables 4 and 5, pp. 249 and 250. N. Higham, *A Very Scientific Gentleman: The Major Achievements of Henry Clifton Sorby*, Oxford, 1963; for example, Sorby, 'On the microscopical structure of

the calcareous grit of the Yorkshire coast', *PYGS*, 1849–59, **iii**, 197–206 (read June 1851).

50 Thomas John Pearsall (1805–83) was then a Leeds consulting chemist and lecturer for the Yorkshire Union of Mechanics' Institutes; James Haywood (d. 1854), was a Sheffield consulting chemist and lecturer.

51 Benjamin Biram (1804–57), coal viewer, who like Hartop worked for Earl Fitzwilliam: Mee, *Aristocratic Enterprize*; A. K. Clayton, 'The Elsecar Collieries under Joshua and Benjamin Biram', unpublished typescript, 1964, Sheffield Public Library.

52 A. H. Green, 'On the coal measures of the neighbourhood of Rotherham', *PYGS*, 1859–68, 4, 685–98 (686–7). cf. Green *et al.*, *Memoirs of the Geological Survey. England and Wales. The Geology of the Yorkshire Coalfield*, London, 1878, pp. 4–7.

53 cf. P. Mathias, 'Who unbound Prometheus? Science and technical change, 1600–1800', in Mathias (ed.), *Science and Society, 1600–1900*, Cambridge, 1973, pp. 54–80; *idem*, 'Science and technology during the industrial revolution: some general problems', *Proceedings of the sixth International Economic History Congress*, Copenhagen, 1978, pp. 104–9.

9 Medical elites, the general practitioner and patient power in Britain during the cholera epidemic of 1831–2

Michael Durey

I

This paper is concerned with the period 1831–2 when Asiatic cholera spread through Britain, killing more than 30,000 people. However, the focus is not on the disease itself, but on how cholera's presence highlighted the structural tensions within professional medicine, the tyranny of public approval of medicine from which the medical community struggled to release itself, and the means whereby medical men sought social status. Two themes are developed. The first relates to the intraprofessional conflicts engendered by the establishment of new, but temporary, institutional structures to combat cholera at the local level. It will be shown that in the provincial towns, open disputes between consultant elites and general practitioners over the medical composition of these boards frequently occurred, which acted against the profession's objective of raising their individual and collective social standing. The second theme is concerned with the relationships between the lay and the medical worlds, and demonstrates that public approval was important in the development of the social image of the medical man. It will be argued that the search for a specific cure for cholera, and the reliance on traditional remedies during the epidemic, can best be understood by reference to the social context in which professional medicine existed.

Both these themes must be seen within the broader framework of a metropolitan–provincial medical dichotomy which existed in the first half of the nineteenth century. In recent years historians have tended to concentrate on this theme, because it offers explanations for the emergence of the general practitioner and for the rise of the modern medical profession.[1] Yet, while metropolitan control of medical education and medical qualifications, at the expense of the less prestigious provincial associations and institutions, is without doubt an important concern for historians, it is arguable that the context of metropolitan–provincial relations can only be fully understood by referring to the wider range of concerns which provincial medical men faced when building their careers. The provincial practitioner not only felt himself circumscribed by the monopolist position of the Royal Colleges in London, he was also constrained by *local* intraprofessional rivalries, and by the needs and demands of his patients. Indeed, this paper suggests that in certain circumstances these local contingencies could be more important to the individual practitioner than any general ideas concerning the role of the medical institutions in London.

Underlying both themes in this paper is the undeniable fact that in the early nineteenth century medicine was not a prestigious occupation. Among the prosperous, contempt for organized medicine was commonplace; and the social successes of the quack St John Long and of the phreno-mesmerist Dr John Elliotson merely confirm society's wider lack of confidence in mainstream medicine.[2] Part of the problem facing the medical profession in its relations with the lay public was caused by the difference between what the profession claimed for itself and what it performed in practice. For all the purported advances in medical science in the early nineteenth century,[3] practitioners still had difficulty in curing even the most trivial complaints. Until the profession could demonstrate qualities superior to quacks, its pretensions remained open to scorn and ridicule.[4] As one popular novelist wrote as late as 1852:

It strikes me forcibly, Tom, that medical science is one of the things that makes little progress, considering all the advantages of our century. I don't mean to say that they don't know better what's inside of you, what your bones are made of, that they haven't more hard names for everything than formerly; but that when it comes to curing you of a toothache, or a colic, or a fit of the gout, my sure belief is they made just as good a hand at it 200 years ago.[5]

Although the profession could claim superiority to both quacks and their predecessors when it came to "heroic" surgery, with the more mundane complaints of life there was little to choose between the quack and the professional. Accordingly, for the more prosperous such as Lady Chettam in *Middlemarch*, a combination of 'constant medical attendances' and a reliance on home remedies became a preferred means of remaining healthy. Such a regimen was not conducive to raising the social position of the medical practitioner, or to fulfilling his rising expectations.

This contempt reflected a tension within community attitudes towards professional medicine, and some distinction should be made between lay feelings about the medical fraternity as a group and lay opinions of individual practitioners. In the first half of the nineteenth century there was an ambivalent lay attitude towards medicine, which distinguished between the personal and the institutional. Thus Lady Chettam could not bring herself to rely exclusively on professional assistance, yet found 'poor Hick's judgement unfailing; I never knew him wrong. He was coarse and butcher-like, but he knew my constitution. It was a loss to me his going off so suddenly'.[6] The collective image of medicine was badly tarnished by its failure to cure, by the publicizing of body-snatching incidents, and by Thomas Wakley's exposures in *The Lancet* of the nepotism, graft and incompetence among hospital staff;[7] but the personal image of the individual medical man remained largely intact. Wakley himself was aware of this ambivalence. In 1832 he asserted that:

The medical profession, *as a body*, command but little of the respect and esteem of the people at large. As individuals, they are admired and beloved within the immediate range of their labours but ... the profession, as a body ... carries not with it the weight and force of national character which belong to the profession of the law or that of divinity.[8]

The doctor–patient relationship, the meeting of the lay world and medical science face to face, is thus a crucial factor in any analysis of the social context within which medicine flourished. In particular it is fundamental to any analysis of the social and professional aspirations of the medical community. This becomes clear when one examines the motivations for membership of the various literary and philosophical and scientific societies which burgeoned in Britain in the early nineteenth century. As has been shown recently, these societies functioned either to reinforce or to legitimize their members' social and intellectual positions in local society.[9] A significant number of medical practitioners belonged to these societies and, as Ian Inkster has argued, they attempted 'to legitimise their social and intellectual positions through social action'.[10]

It can also be argued that for many in the profession, membership of societies helped to raise their status rather than just to legitimize the social positions they already possessed. It seems inherently more likely that, for the majority of general practitioners at this time, the desire for upward social mobility was more pertinent than a need to confirm status. It has been estimated in 1848 that there were between 14,000 and 15,000 general practitioners in England and Wales.[11] The vast bulk of these would have found it particularly difficult to build up a numerous clientele, of a social composition which guaranteed adequate remuneration.[12] Competition between practitioners for scarce resources could be fierce and any advantage which could be gained over one's rivals was eagerly seized.

Membership of local scientific societies was one way in which the social position of the general practitioner could be raised. Such membership enabled him to share in the cultural life of his community, offered a prestigious social image to the world, and perhaps most important, gave access to fellow lay members who were potential patients. However, the individual medical man was concerned not only with his own advancement, but also with the status of his profession; and while it seems clear that membership of local cultural societies raised the prestige of medical individuals within the community, it is by no means so obvious that the medical profession as an occupation shared in this glory. For middle-class manufacturers or lawyers, a membership card to the local scientific institution led to a recognition, in social and cultural terms, of themselves as individuals. But for the medical man, this was insufficient: his ultimate status depended on the professional authority of his occupation. The essential difference between the manufacturer and the medical man was that the former did not require the collective social image of his caste to be tested by public opinion. The social image of the medical man, however, could be scrutinized; and the forum was the consultation, which was personal, direct and unequivocal. Thus, what medical men actually did in their capacity as medical practitioners enabled the lay public to endorse or reject the image presented by membership of scientific societies.

It is clear, therefore, that any recognizable success in the field of medical science would confer greater prestige on the medical profession as a whole than individual practitioner's membership of Mechanics' Institutes or scientific societies, for it would confirm the utilitarian function of medicine. Public acceptance of medicine as a high status occupation depended on the profession's ability to persuade lay opinion that it had mastery over its subject, was responsible and could respond to emergencies. The appearance of cholera for the first time on British soil in October 1831 gave the medical community an opportunity to respond to a crisis, to demonstrate professional medicine's utility, and, as a consequence, to raise both individually and *collectively* the prestige of their profession. The working out of these opportunities in practice is the subject of the following sections.

II

In 1831 Britain was almost totally unequipped to cope with an invasion by a new and potentially virulent disease. A centralized system of public health did not exist. Apart from a small group of Benthamites, there was little adherence to the concept of a medical police, that is the creation of a medical policy by government and its implementation through government regulation. The only state medical bodies were the quarantine establishment, the Commission in Lunacy and the Vaccination Board, all of which were supervised by the Privy Council.[13] Thus the quarantine system, established in the early eighteenth century to prevent contagious or infectious diseases from entering Britain, was the only means of defence against an epidemic from abroad. Penetration of this curtain, which in 1831 was delapidated through underuse,[14] left the country totally unguarded.

With no national system of preventative medicine available, the second line of defence against an epidemic was the medical profession which in 1831 was demoralized and in some disarray.[15] During the first decades of the nineteenth century the twin processes of industrialization and urbanization had undermined the antediluvian structure of the profession. The diffusion of wealth among a growing middle class had increased the pool of potential patients capable of paying doctors' fees; their living in or near the burgeoning industrial towns of the Midlands and North had placed them in areas outside the control of the two major medical institutions, the Royal College of Physicians and the Royal College of Surgeons. By 1830 a large proportion of the middle class had access to doctors, whose qualifications and experience varied greatly. The old tripartite division of medical men into physicians, surgeons and apothecaries, each with their own specialist functions, no longer conformed to reality. Instead, even in London, there was emerging a new breed of doctor, the general practitioner, able and willing to prescribe medicines as well as to act as surgeon, man-midwife and pharmacist.[16]

Such a development naturally caused tensions within the medical community. The old structure was based on a strict hierarchy of orders into which it was impossible to fit the general practitioner, for he performed the functions of all three types of medical man. At the apex of the medical community were the physicians, particularly the Fellows of the Royal College, who fought hard to maintain their privileges and their high social status, which, claimed one, was a consequence of their sharing an exclusive education with the gentry at Oxford and Cambridge.[17] Holloway has argued, convincingly, that by supporting the Apothecaries' Act of 1815 the Fellows managed successfully to confirm their status and to defuse the medical reform movements which originated among the junior ranks of the profession early in the century. Although the Act increased the Society of Apothecaries' jurisdiction and augmented its powers, it also kept the hierarchy of orders intact and emphasized the low-status trade origins of the Society.[18] In essence the Act worked to block the social pretensions of both provincial and metropolitan general practitioners who eagerly sought a share of the growing middle-class market. But within the provincial medical community there was also a widening social gap between the physicians, who usually had been educated either at Oxford or Cambridge or at one of the Scottish universities and who held honorary posts in the provincial public institutions, and a lower stratum of general practitioners who found themselves, as did their metropolitan counterparts, blocked by a medical elite from achieving a higher social position. This division *within* provincial medical society is crucial in understanding responses to cholera in 1832, and metropolitan–provincial relations.

Cholera first became an issue in the provinces with the publication on 20 October 1831 of an Order in Council, which called on local authorities to establish boards of health. The Order had developed from recommendations given to the Privy Council by the General Board of Health, which was completely dominated by Fellows of the Royal College of Physicians and was under the chairmanship of Sir Henry Halford, the College's President. The Privy Council, mindful of its lack of powers of intervention in local affairs, hoped that these boards would be the best way of harnessing local pride and initiative in the war against cholera. Traditionalist and localist orthodoxies concerning the role of the State were revealed in a three-tier system consisting of the Central Board, district boards (covering a whole town) and subordinate divisional boards (covering individual parishes within a town). All sections of the ruling class were expected to be represented on the district board, while the divisional boards were to be composed of resident clergymen, a number of substantial householders and at least one member of the medical profession.

Emanating from a Board of Health monopolized by the Royal College of Physicians, it is not surprising that in this scheme the role envisaged for the provincial general practioner was minimal. The social status of medical practitioners in each concentric ring of the system was to conform to the status of their fellow lay members: medical expertise or knowledge of working-class conditions was seemingly irrelevant. Understandably, Thomas Wakley considered the proposals as 'little short of a premeditated deliberate insult, offered to a majority of general practitioners . . .'.[19] Every medical man in each district, he argued, should be on the board and together they should have entire control of public health affairs. In this instance, no metropolitan–provincial clash occurred: indeed, eight out of every ten parishes eventually established local boards of health, under the guidelines of the Privy Council. Accommodation was reached because the Privy Council rarely tried to interfere in local affairs during the epidemic, even when the Cholera Prevention Act of February 1832 gave it in theory unlimited powers in public health matters. It was the dismissal of Halford's Board of Health in November 1831 and its replacement by one on which Sir David Barry and Sir William Russell, both Edinburgh graduates, were the medical representatives, which conciliated provincial medical opinion.[20] Recommendations from London were no longer seen as "commands" from the Royal College of Physicians, and metropolitan–provincial relations remained firm.

Nevertheless, Wakley's concern with the possibilities opened up to local medical monopolists by the establishment of local boards of health was well founded, especially in the larger provincial towns.[21] The most prestigious members of the local medical communities, the licentiates of the Royal Colleges and the honorary consultants at the voluntary hospitals, who frequently had close connections with the local ruling elites, predictably took the major medical positions on the district boards of health.[22] In York the Central Committee included all the city's physicians and the six senior surgeons, who combined to become a General Medical Board.[23] In Worcester, medical representation on the Board of Health consisted of only a few physicians, including Charles Hastings, who did not bother to inform the rest of the medical community of their new posts.[24]

This accretion of influence to local medical elites did not always occur without an internal struggle. In November 1831, for instance, the Liverpool magistrates appointed a committee of five physicians to advise the Board of Health. All were "friends and protégés" of the magistrates; all were attached to the local public charities. Circulars were then sent to the rest of the medical community, asking them to attend a meeting at which they would be allocated a particular district of the town. At the meeting, attended by 100 practitioners, vigorous objections were made to the means of selection and the composition of the medical committee. The public, claimed Dr Carson, had no confidence in these men. The public knew

That the officers of those charities were not appointed to them in consequence of superior talents or greater claim to respectability; they knew indeed that their superiors, certainly their equals, were among those who had *not* so been appointed, but who, from their years and acknowledged talents, ought to have been allowed to give the benefit of their abilities to the public charities.[25]

Too much influence, he alleged, was being concentrated in too few hands. For those appointed to the public institutions, 'the honour was too often the result of their disposition to pander to the base passions and prejudices of those whom they thought could shed the influence of their local patronage or petty authority on them'. Already they monopolized most of the influential positions in Liverpool, and by refusing to give way they operated 'like an incubus on the zeal and prospects of their younger brethren'.

A motion calling for a new committee, elected from among the body of the profession as a whole, seemed certain to be passed; but the chairman, Dr Renwick, himself one of the five under attack refused to put the resolution and walked out of the hall. His successor then put the motion confirming the appointment of the five, but only twelve voted in favour. An open vote was then called for; but, forced into a position where 'they might be made the objects of the odium and persecution of the aristocratic professionals and non-professionals of the town should they evince liberality of feeling', only a few had the courage to risk the loss of their clientele by opposing the motion.[26] Thus a small medical elite, supported by the local ruling class, had its way. The bulk of the medical community faced a terrible dilemma; dependent on the local middle-class merchants and shipowners for their clientele, they could not risk offending that clientele by rejecting its selection of local board members.

It would be a mistake however to accept too readily Wakley's statement that most towns and cities in the country demonstrated the type of intraprofessional conflict which occurred in Liverpool. It was possible in fact for a majority of the profession to inflict considerable damage on the board of health, but only if the board did not represent a majority of the public or if its authority was weak. In the Gorbals, Glasgow, a small group, chosen by the Glasgow magistrates, formed the local board of health, on which sat just three medical men. Public opinion was very hostile to this junta, for it not only failed to make any sanitary arrangements, but also misused private subscriptions raised for cholera relief work and failed to open a temporary cholera hospital which had been paid for by private donations. It became possible, when its offers of help had been ignored, for the local medical community 'in a body [to] draw off, and publish their secession from all connections with the establishment'. They had no fears of losing their patients, for 'The medical practitioners paid by the board are never sought for; while others of known and respectable standing are regularly applied to.'[27]

Some preliminary grumbling about the medical composition of the district boards of health may have been common,[28] but usually this could be overcome by the judicious use of patronage. In Worcester, a public meeting called to adopt resolutions from the Board of Health made the local medical men aware that they had not been consulted 'by their learned and dictatorial colleagues'. The Board of Health met the twenty-six general practitioners and received offers of gratuitous assistance once the serious situation had been explained. Eleven medical men, however, later passed a resolution expressing dissent 'from the principle of exclusive appointment'. Their opposition was to be stifled when each parish was put under the control of two or more medical officers.[29]

By giving each medical man a slice of the cake, the positions of the elites on the prestigious district boards of health could usually be confirmed and open conflict avoided. In Edinburgh, where 250 medical men practised, the district board had fourteen of its twenty-seven positions filled by doctors. Naturally, several famous physicians and surgeons were included, but so also were five ex-Indian army general practitioners. Moreover, the city was divided into thirty districts, and 100 medical men formed an association to visit them. All were of the same rank and no disunity occurred.[30] In Sheffield, the numbers of medical practitioners on the district board fluctuated according to the gravity of the situation.[31]

Wakley's judgement was thus made too early in the saga of the local boards of health, before he could see how they were to work in practice. He failed to realize that, possibly for the first time, most members of each medical community were given some official recognition. Admittedly, while the system certainly did not place all practitioners on the same social footing, it did encourage interplay between the different status groups. Moreover, the local board structure, by placing many practitioners on the same divisional committees as local clergymen, overseers and principal inhabitants, helped to raise the status of the apothecary and general practitioner *at the parish level*. For many of the profession this would have been sufficiently attractive to ensure that they did not question the distribution of influence at the city level.

It is also worthwhile pointing out that by no means all practitioners in Britain found themselves in competition with each other. In the smaller communities, and especially in the rural areas, there may only have been two qualified practitioners available for selection to the local board of health. When cholera broke out in Goole, in April 1832, there was only one qualified man in practice there.[32] The nearest practitioner to the township of Shields when cholera broke out in November 1831 lived five miles away. One of those who left Newcastle to give assistance was a medical student called John Snow.[33] In the huge rural parish of Plympton in Devon, Mr Langworthy, the sole representative of qualified medicine, was fortunate to find an able assistant in the Reverend Coppard, the local incumbent.[34]

Thus the opportunities offered by the establishment of a voluntary system of local boards of health differed from area to area. Without doubt this new form of health administration offered new ways for the medical elites to cement their pre-existing linkages with the local magistracy. For the ordinary general practitioner in the growing middle-class conglomerates of the Midlands and North of England, some prestige was attached to membership of divisional boards of health. It would not have helped those with aspirations to advancement in the public institutions, but it may have brought some additional clientele. As many medical practitioners lived on the margin of subsistence, an increase in the number of patients, even if they were local butchers and greengrocers rather than important merchants and industrialists, must have been a boon. In the more sparsely populated rural areas and small market towns, membership of the board of health was automatic. There was less prestige involved, but there were fewer to share it. If nothing else, that hazy division between the qualified and the unqualified would have been temporarily widened.

But there was occasionally a considerable price to pay for this prestige. Once the local boards were established with their compliant medical officers, they were sometimes used to defend local capitalist interests against the threat imposed by central regulations, especially those concerning quarantine. The most frequent tactic was to deny the presence of cholera in the district, claiming either reduced mortality for that time of year or the presence of indigenous English cholera (summer diarrhoea).[35] The role required of the medical community was to buttress such claims, but occasionally sections of the profession had to be bludgeoned into submission. The most notorious example occurred in the autumn of 1831 in Sunderland, the first town to be infected by cholera.

The attempted cover-up by the local ruling classes and the suborning of the local medical profession only came to light because of the fortuitous presence in Sunderland of the military surgeon James Butler Kell, who not only had witnessed cholera in Mauritius in 1829 but also personally knew Sir James McGrigor, then on Halford's Board of Health. Kell was very conscientious and soon became suspicious when four isolated cases of a disease similar to cholera occurred in August.[36] No official reports were made of these; it was only when Kell attended a meeting of the local Board of Health (of which he was an honorary member) on 30 August that he learned of the death of one Robert Henry, a pilot.[37] Inquiries having satisfied him that Henry had piloted a ship from the Baltic into harbour shortly before his death, Kell immediately reported his fears to McGrigor in London without informing the Sunderland authorities.[38]

Kell's independent action and his subsequent complaints about the quarantine establishment angered the local authorities; but, primarily because of the few and isolated cases at this juncture, they were able to refute the surgeon's claims and silence the London Board of Health's inquiries with relative ease.[39] Five weeks elapsed before Kell heard of further cases, the first of which occurred close to where several ships were completing quarantine. Soon afterwards he learned, again by chance, of William Sproat Snr's death, which was subsequently to be accepted as the first cholera fatality in the British Isles. Kell once again informed London, although he later claimed to have asked a local clergyman to persuade George Robinson, Chairman of the local Board of Health, to write to London. No message was received from Robinson.[40]

Dr Reid Clanny, Chairman of the medical subcommittee of the Board of Health and physician to the local infirmary, was also convinced of cholera's presence but he failed to obtain the agreement of his medical colleagues on the board. On 1 November, however, after several more fatal cases had been reported, a general meeting of the medical fraternity reversed this decision with no dissenting voice. The local board reported this decision officially to London.[41] When only one further case was reported in the next few days (the spasmodic incidence of cholera at the beginning of an epidemic is a common phenomenon), doubts were expressed about the wisdom of the 1 November decision, especially as the implementation of quarantine regulations was certain to have unpopular economic repercussions. Of the thirty-five medical men practising in Sunderland, twenty-eight lived in the middle-class parish of Bishopwearmouth and were dependent on the good offices of their middle-class patients. They began to rue their hasty decision of 1 November.

When the three government agents, Robert Daun, Lieutenant Colonel Michael Creagh and Dr Gibson, arrived in Sunderland, they faced a growing conspiracy of silence. Daun was soon at his wit's end. He could only confirm the presence of Asiatic cholera if he were given access to the victims, but professional etiquette prevented him from visiting the sick without permission from the attending doctor. Several times Daun waited in vain for a doctor to arrive. At first he charitably excused their non-appearance on the grounds of overwork, but he quickly realized the real reason for this obstructionism. 'It had been signified to [me] . . . that the mercantile and shipping interest of this port, dreading that restriction might be imposed upon the commerce, are strongly opposed to the belief that cholera exists here in any unusual degree or form and that they are supported in this view of the case by not a few of the medical practitioners here.'[42]

The real onslaught by the commercial classes came in the second week of November, when at a series of meetings two resolutions were passed, one condemning the erroneous reports sent to London and the other in favour of publicly naming those doctors present at the 1 November meeting. This direct threat of intimidation was immediately successful. Of the nineteen medical practitioners who had met that day to express their opinions on cholera, only one was prepared to defy the commercial interests by claiming the existence of Asiatic cholera. At least four of these doctors had previously attended cholera cases, and one had even suffered from the disease.[43] The rout was completed the next day when twenty-eight medical men signed a statement which, while admitting the presence of a disease with every sympton of epidemic cholera, stated that it was not contagious, had not been imported, and would only be worsened by the pernicious effects of quarantine regulations.[44] Even Clanny signed the document, thus isolating Kell, although he had the excuse of having had his life threatened. Daun had informed the Central Board on the previous day that a mob of sailors was going to attack Clanny's house that evening.[45]

At the same time as the medical profession was wilting before the economic facts of life, the commercial and shipping interests turned their attention to the local Board of Health. At yet another meeting on the 11 November, resolutions hostile to Kell, Clanny and George Robinson were passed and the Local Board was replaced by one more amenable to the pressure of specific interest groups. Daun informed London on the 15th that 'a new board of health has been constituted consisting of God knows how many members – a great proportion of them shipowners and merchants – on the ground forsooth that their interests must be represented on the board!!!'.[46]

Thus, as in Liverpool, a specific interest group within the local ruling class found it comparatively easy to steamroller the dependent medical profession into submission. Similar pressures were applied to the Newcastle medical community when cholera broke out in November 1831. On this occasion the Mayor on his own initiative reported the first fatal case to London, an action with which the local Board of Health violently disagreed. Newcastle was placed under an immediate ten-day quarantine, which even a unanimous resolution from the medical profession denying cholera's existence failed to rescind. Sufficient pressure was put on the Mayor for him to recant, but within four days further cases occurred which put the issue beyond doubt.[47]

The denial of cholera was to be a major tool for those whose profits were threatened by government regulation in 1831–2. The acquiescence of the medical profession was vital for the success of this policy; a show of independence could not be tolerated. In

December 1831, the Manchester Board of Health, no doubt mindful of the experiences in the North-East, ordered that no medical man should make public any case of cholera until the medical board had given its opinion.[48] In London, where the shipowners had called meetings of protest against the quarantine laws in June 1831, petty shopkeepers and artisans used their influence on local vestries to protect their personal interests. A reluctance to sanction a new rate and the fear that potential customers might be scared away were two factors influencing the shopocracy's viewpoint. At least one parish refused to establish a board of health, and in others the boards issued statements to the press denying the existence of cholera.[49] On every occasion the medical members of the board would have been an important instrument, masking pecuniary interests with scientific expertise.

III

Pressures on the medical community to use its scientific standing for the benefit of local economic interests were most acute in seaport towns and cities which were merchant-intensive, and when cholera's presence was problematic. Once considerable numbers fell victim to the disease, the pretence of denial could not be sustained. Merchants' and shipowners' concern for their profit margins declined when, following London's infection in February 1832, the Government abandoned quarantine regulations. Thereafter, pressures on the medical communities were felt only when local shop owners and artisans, realizing that the presence of cholera could be disastrous for trade, had sufficient influence in the vestry to sway the opinions of local medical men.[50] Again, however, once the disease was firmly entrenched in a community, denial became a useless tactic. This left the medical community free to use their expertise unimpeded. On the results of their therapy depended their future prestige, both collectively and individually.

It should be stressed that specific remedies were extremely rare at this time. Most drugs available were merely palliative, at best diminishing the side-effects of a disease without destroying the pathogenic micro-organism. Some new drugs such as morphine, quinine and, in 1832, codeine had been isolated, but they had not been fully tested, nor had they been included in the most recent pharmacopoeia. Indeed, one London medical journal commented in 1832 that 'It is passing strange that our *Pharmacopoeia* should always be behind the progress of science.'[51] Lack of available drugs severely limited the range of choice open to the medical profession when facing cholera in 1832. Usually practitioners depended on a series of drugs and techniques to treat each set of symptoms as they arose. Most frequent treatments were blood-letting at a very early stage of the disease, followed by calomel and opium in various combinations. From the perspective of contemporary scientific theory, there were good reasons for relying on these particular traditional modes of treatment.[52]

But pressures to remain conservative, and to be very suspicious of any new therapy, were also exerted by the practitioners' concern for their reputation among the lay public. Any prolonged testing of a new therapy which failed to show successful clinical results endangered whatever social standing and respect the practitioner had attained. This applied whether he was a London physician, a member of an upwardly mobile group in the provinces, or an isolated practitioner in the countryside. One result of this was the failure of intravenous saline injections, which were first suggested by the surgeon William O'Shaughnessy late in 1831,[53] and which, in favourable clinical conditions, are still the best means of curing a cholera patient. Experiments were carried out throughout the country in 1832, but, after variable initial success, they were abandoned.[54]

The failure of intravenous saline injections in 1832 should not be seen in isolation. Other new therapies, such as croton oil and cauterization (both extremely dangerous), were also discarded by the profession after a minimum of experimentation. It is important to recognize that it was the same rigorous demand for empirical confirmation of theory which led to the demise of *both* potentially useful *and* potentially dangerous therapies. With the possibility of experimentation taking place only in the homes of the sick or in temporary cholera hospitals which were very unpopular, practitioners could not risk losing their professional standing by testing new therapies on their patients.[55]

Indeed, the practitioner had much to gain by displaying reasoned judgement and calmness at the bedside; in a period of crisis the relatives of a victim would be expecting reassurance and comfort from a medical man, not experimentation. It was certainly the case that in Plympton, relatives of the sick awaited the arrival of Surgeon Langworthy with tense anticipation.[56] Nor did he let them down. Almost dropping from fatigue, he rode across his vast practice doing what little he could to alleviate suffering. He was by no means a hick doctor: he read about intravenous saline infusions in *The Lancet* and tested them, but the need to visit other patients prevented him giving the system a fair trial. There is no doubt that Langworthy's exertions, and the help given by the local incumbent, gave a great psychological boost to both patients and relatives. Indeed, one does not know whether he assuaged fear and anxiety not just by being present, but precisely *because* he used traditional remedies. Janet Blackman has shown how long medical knowledge could take to percolate down the social scale and how it could be moulded to fit existing popular perceptions.[57] Blood-letting might well have been tolerable;[58] newer techniques, especially the external therapies such as cauterization and the cold water treatment (the latter suggesting the treatment given to the insane), probably were not. Thus, by keeping to old and tested remedies, the medical community might have reinforced popular perceptions, reduced suspicion of official medicine and raised or at least confirmed the social position of the individual practitioner. Much could have been lost by a frivolous regard for innovation for its own sake.

Not every practitioner in 1832, however, preferred to pursue a safety-first policy. There were some who saw the epidemic as a means of promoting their own personal status, both within the medical and the local communities, by claiming to have discovered specific cures. Prominent among these was Dr William Stevens, who had first suggested his saline plan (the infusion orally and per rectum of a solution of neutral salts) to treat yellow fever patients in Tobago in the late 1820s.[59] Stevens exemplified those medical men who used a specific therapy as a way of gaining both fame and an entrance into the relatively closed world of the medical establishment. His treatment was at first taken extremely seriously and the Central Board of Health recommended its testing. But in June 1832 Stevens claimed the cure of ninety-two cholera victims in Cold Bath Fields House of Correction in London. A Central Board medical team found that none of the patients had been suffering from cholera; one was diagnosed as a suitable case for a dentist.[60] Stevens's reputation was destroyed; the combination of unsuccessful clinical results and the social disgrace of its mentor ensured that the saline therapy was totally discredited, although variants of it were to be used in later epidemics.

In the provinces, the desire to extol one's own particular

remedies can be understood by reference to the competition from quacks whose proprietary medicines were extremely popular.[61] Those who succumbed to the temptation of competing in the market-place with charlatans predictably found themselves involved in intraprofessional disputes. Clashes occurred in Manchester and Liverpool,[62] but the most colourful happened in Hull, involving Dr Joseph Ayres, which culminated in a challenge to a duel with the vanquished to leave town.[63] Ayres was the true culprit, a provincial practitioner seeking prestige by claiming a specific cure for cholera which was 'not merely successful, but eminently successful'.[64] His specific was not original, being merely vast quantities of calomel and opium. Ayres was a persistent character and was to be found as late as the 1850s hawking his cures around the medical journals.[65] It was one of Ayres's opponents, John Alderson, who most clearly pointed to the alternative approaches which a practitioner could take in 1832. One could either seek a specific remedy, or, 'by observation and comparison', establish a rational plan of treatment.[66] Alderson was firmly in favour of the second approach: only those seeking 'easy public success' would ever claim to have discovered a specific cure for cholera. Practitioners such as Stevens and Ayres were very much in a minority in 1832, most of their comrades preferring to follow Alderson in seeking a rational plan of treatment. But Ayres received much more national publicity than Alderson. As a result, historians have tended to accentuate the importance of the mavericks in the medical community and to neglect the majority, who perceived that the desire to relieve suffering, and the need to gain status both in the eyes of their peers and of lay society, usually dovetailed to suggest that cholera should be treated in a traditional, swift and "safe" manner.

IV

The medical community faced the cholera epidemic in 1831 in chaos and with morale at a low ebb. Within months its major institution, the Royal College of Physicians, had been humiliated; and in several provincial towns medical practitioners had been made aware, with varying degrees of subtlety, of the power and influence of the lay patient. Then the disease had to be combated, not only with a weak and narrow range of resources but also with one eye on how the relatives of the sick would respond to the application of experimental therapies. It is a measure of the resilience of the medical community that most practitioners combated cholera manfully, with both compassion and fortitude. Some died treating the sick; many suffered from overwork and overstrain.[67] In some areas medals were struck for practitioners as a mark of respect for their endeavours during the cholera epidemic. There can be little doubt that many individual practitioners achieved considerable success in promoting their own social status in 1832.

Nevertheless, the medical profession *per se* did not emerge from the epidemic with increased prestige. This was not owing to any conspiracy among London medical elites aimed at restraining provincial practitioners; indeed, the lines of communication between London and the provinces remained open in 1832 and there is impressive evidence of the wide dispersion of medical information and much fruitful co-operation throughout Britain.[68] The credit for this must lie with Russell and Barry in London, who ensured that medical journals were given up-to-date information and that only recommendations, not orders, were sent to the provinces.

The explanation for the profession's unchanged position following the epidemic must be sought at the *local* level, and especially in the contradiction between the social image displayed by the medical profession and the reality of their achievements at the bedside. The strength of the practitioners' positions lay in the sympathetic care they gave to the sick in their homes; but this patient–doctor relationship was not publicly visible, being observed at first-hand only by the victim's relatives and, possibly, neighbours. For those of the public who did not become infected, especially among the middle classes who were not the disease's primary victims, the medical profession's image depended on its public activities; and here there were too many examples of weak behaviour at the local level for the profession to succeed in convincing the laity of its scientific and moral authority. The inconclusive debates over etiology, which were first heard at the meetings of local medical societies, but which later spilled over into the newspapers,[69] the refusal of the London consultants, hiding behind a strict interpretation of the rules and regulations concerning infectious diseases, to allow cholera victims entry to their hospitals, the obvious dissensions within local medical communities over appointments to the new boards of health, and the priority disputes over therapy, all combined to reduce rather than increase the status of the profession. The collective image of medicine was one of intraprofessional squabbling and weakness when confronted by external pressure. Not until the profession could demonstrate that it could control its own local activities and could keep its own house in order, would its social prestige and influence rise.

Notes and references

1   S. W. F. Holloway, 'The Apothecaries' Act, 1815: a reinterpretation', *Medical History*, 1966, **10**, 107–29, 221–36; *idem*, 'Medical education in England, 1830–58: a sociological analysis', *History*, 1964, **9**, 299–324; N. Parry and J. Parry, *The Rise of the Medical Profession: A Study of Collective Social Mobility*, London, 1976; I. Waddington, 'General practitioners and consultants in early nineteenth-century England: the sociology of an intra-professional conflict', in J. Woodward and D. Richards (eds.), *Health Care and Popular Medicine in Nineteenth Century England*, London, 1977, pp. 164–88; M. Jeane Peterson, *The Medical Profession in Mid-Victorian England*, Berkeley and Los Angeles, 1978.

2   *The Lancet*, 1831, **1**, 296; for John Elliotson, see DNB.

3   *Edinburgh Medical and Surgical Journal*, 1832, **37**, Appendix, cclxx.

4   R. Sand, *The Advance to Social Medicine*, London, 1952, p. 56.

5   Quoted in M. F. Brightfield, 'The medical profession in early Victorian England, as depicted in the novels of the period 1840–70', *Bulletin of the History of Medicine*, 1961, **25**, 253.

6   G. Eliot, *Middlemarch*, ch. 10.

7   S. S. Sprigge, *The Life and Times of Thomas Wakley*, London, 1899; C. Brook, *Battling Surgeon*, Glasgow, 1945.

8   *The Lancet*, 1831–2, **2**, 88.

9   A. Thackray, 'Natural knowledge in cultural context: the Manchester model', *American Historical Review*, 1974, **79**, 672–709.

10 I. Inkster, 'Marginal men: aspects of the social role of the medical community in Sheffield, 1790–1850', in Woodward and Richards, op. cit. (1), p. 129.

11 Waddington, op. cit. (1), p. 168.

12 Peterson, op. cit. (1), ch. 3.

13 C. F. Brockington, *Public Health in the Nineteenth Century*, Edinburgh, 1965, p. 1.

14 J. Marshall to Sir William Pym, 6 and 9 November 1831, PRO PC1/4395; *The Lancet*, 1831–2, **1**, 669–70; *Edinburgh Medical and Surgical Journal* 1831, **37**, 202; *The Times*, 5 November 1831.

15 For contemporary confirmation see for example, *London Medical*

and Surgical Journal, 1832, **1**, 17; C. Hastings, 'An address delivered at the first meeting of the Provincial Medical and Surgical Association', *Transactions of the Provincial Medical and Surgical Association*, 1833, **1**, 5–6.

16   Waddington, op. cit. (1), p. 166.

17   Select Committee on Medical Education, PP, 1834, **13**, p. 35 (evidence of William Macmichael).

18   Holloway, op. cit. (1).

19   *The Lancet*, 1830–1, **2**, 820.

20   For a fuller analysis of Halford's Board of Health see M. Durey, *The Return of the Plague: British Society and the Cholera*, Dublin 1979, ch. 1.

21   *The Lancet*, 1831–2, **1**, 305.

22   ibid, 305.

23   *Yorkshire Gazette*, 3 December 1831.

24   W. H. McMenemy, *The Life and Times of Sir Charles Hastings*, Edinburgh, 1959, p. 78.

25   *The Lancet*, 1831–2, **1**, 315.

26   ibid, 308.

27   *Glasgow Chronicle*, March 1832. A copy of the letter from David Laurie can also be found in PRO PC1/103.

28   The Lancet, 1831–2, **1**, 315.

29   McMenemy, op. cit. (24), p. 79.

30   *Edinburgh Medical and Surgical Journal*, 1832, **37**, Appendix, cclxx.

31   J. Stokes, *History of the Epidemic Cholera in Sheffield in 1832*, Sheffield, 1921, pp. 42–3.

32   E. Denison to Central Board of Health, 11 April 1832, PRO PC1/105.

33   B. W. Richardson, 'John Snow, M.D.', in W. H. Frost (ed.), *Snow on Cholera: Being a Reprint of Two Papers*, New York and London, 1965, p. xxvi.

34   W. I. Coppard, *Cottage Scenes during the Cholera*, London, 1848.

35   Durey, op. cit. (20), ch. 6.

36   W. R. Clanny, *Hyperanthraxis: Or the Cholera in Sunderland*, London, 1832, p. 13; *London Medical Gazette*, 4 February 1832.

37   J. B. Kell, *On the Appearance of Cholera at Sunderland in 1831*, Edinburgh, 1834, pp. 21–3.

38   ibid, p. 24; Kell to McGrigor, 2 and 3 September 1831, PRO PC1/4395.

39   Kell, op. cit. (37), pp. 27–9; Kell to McGrigor, 9 November 1831, PRO PC1/4395.

40 Clanny, op. cit. (36), pp. 13–20; Kell, op. cit. (37), pp. 30–3; Kell to McGrigor, 26 and 27 October 1831, PRO PC1/4395; *The Times*, 5 November 1831.

41 Clanny, op. cit. (36), p. 42; Kell, op. cit. (37), p. 36; minutes of the Central Board of Health, 3 November 1831, PRO PC1/105.

42 Kell, op. cit. (37), pp. 39–40; Daun to Privy Council, 8 and 9 November 1831, PRO HO 31/17.

43 *Sunderland Herald*, 12 November 1831.

44 Kell, op. cit. (37), p. 49.

45 Daun to Privy Council, 11 November 1831, PRO PC1/4395.

46 Resolutions of merchants and shipowners, 11 November 1831, PRO PC1/4395; Daun to Privy Council, 15 November 1831, PRO PC1/103.

47 Minutes of the Central Board of Health, 29 November, 5 December 1831, PRO PC1/105; *Newcastle Chronicle*, 3 December 1831.

48 Proceedings of the Manchester Board of Health, 21 December 1831, Manchester Public Library, M9/36/1.

49 Vestry minutes, Christchurch, Surrey, 22 March, 5 April 1832, Greater London Record Office, P92/CTC/88; Vestry minutes, St Paul's Covent Garden, 23 February 1832, Westminster Public Libraries, H 808.

50 Vestry minutes, Greenwich, 2 August 1832, Greenwich Public Libraries; minutes of Central Board of Health, 29 March 1832, PRO PC1/105; *Leeds Mercury*, 15 September 1832.

51 *London Medical and Physical Journal*, 1832, **8**, 14.

52 Durey, op. cit. (20), ch. 5.

53 W. B. O'Shaughnessy, 'Proposal of a new method of treating the blue epidemic cholera by the injection of highly-oxygenised salts into the venous system', *The Lancet*, 1831–2, **1**, 366–71; *idem, Report on the Chemical Pathology of Malignant Cholera*, London, 1832.

54 Coppard, op. cit. (34), p. 109; J. Alderson, *A Brief Outline of the History and Progress of Cholera at Hull*, London, 1832, pp. 36–7; H. Gaulter, *The Origins and Progress of the Malignant Cholera in Manchester*, London, 1833, p. 148; J. P. Needham, *Facts and Observations on the Cholera in York*, York, 1832.

55 Durey, op. cit. (20), ch. 7.

56 Coppard, op. cit. (34), *passim*.

57 J. Blackman, 'Popular theories of generation: the evolution of *Aristotle's Works*', Woodward and Richards (eds.), op. cit. (1), pp. 56–88.

58 P. H. Niebyl, 'The English bloodletting revolution, or modern

medicine before 1850', *Bulletin of the History of Medicine*, 1977, **51**, 469–70.

59   W. B. Stevens, 'On the efficiency of saline agents in the treatment of West Indian fevers', *The Lancet*, 1831–2, **1**, 553–65.

60   ibid., 592; 1831–2, **2**, 455–7.

61   *Leeds Mercury*, 14 and 28 July 1832.

62   Proceedings of Manchester Board of Health, 28 March 1832; *Liverpool Courier*, 7, 14 and 28 March 1832.

63   *Hull Portfolio*, 12 June 1832.

64   *The Lancet*, 1831–2, **1**, 18.

65   N. Howard-Jones, 'Cholera therapy in the nineteenth century', *Journal of the History of Medicine*, 1972, **28**, 380.

66   Alderson, op. cit. (54), p. 2; *The Lancet*, 1831–2, **2**, 19–20.

67   *The Times*, 6 February 1832; minutes of St George Hanover Square Board of Health, 23 July 1832, Westminster Public Libraries, C 971. Isolated examples of medical men resigning from local boards of health because they feared loss of clientele can be found in *Liverpool Courier*, 20 June 1832, and Gaulter, op. cit. (54), p. 191; and a case of a workhouse doctor experimenting on his patients can be found in Sharman to Central Board of Health, 20 September 1832, PRO PC1/102.

68   Lewis to Maclean, 18 May 1832, PRO PC1/103; *The Lancet*, 1831–2, **2**, 594–5; Anderson to Barry, 21 June 1832, PRO PC1/103; Durey, op. cit. (20), ch. 5.

69   Durey, op. cit. (20), ch. 5.

General index

Adelaide Gallery, 98, 110
Annals of Philosophy, 107
Apothecaries Act (1815), 135, 262
Architectural Society (London),
    100
Ashmolean Museum, 193
Askesian Society (London), 15, 95,
    121–2, 131–7, 143
Astronomer Royal, 92
astronomy, 13
Athenaeum, The, 75, 76, 79

Bath, 180, 182
Beck Affair, 72–5
Benthamites, 58, 136, 261
biology lectures, 104–7
Birmingham, 29, 38
blood-letting, 272
Bradford, 27, 31, 222, 233, 244, 248
Bradford Philosophical Society, 234
Bristol, 20–1, 27, 30, 34, 179–204
Bristol College (1830–41), 194, 196,
    203
Bristol Institution (1823), 21, 29,
    181, 185–91
Bristol Library, 187
Bristol Literary and Philosophical
    Society (1809), 184
Bristol Mechanics' Institute, 191–2
Bristol Medical School, 195
Bristol Philosophical and Literary
    Society, 186–7, 189–91
Bristol Society of Enquirers, 192,
    202
Bristol Statistical Society, 194–5,
    203
British Association for the
    Advancement of Science, 15, 38,
    56, 58, 66–7, 68–71, 73–87, 115,
    158–60, 180–1, 191, 193–4, 206,
    235, 242–4
British economy, 44–5
British Mineralogical Society, 25,
    95, 120–50

catastrophism, 189–90
Central Board of Health, 262–3,
    267–9, 277
Chapter Coffee House Society, 123
Chesterfield, 28
cholera (1831–2), 67, 108, 257–78
Cholera Prevention Act (1832), 263
civic science, 35, 52
Civil List pensions, 68, 92
clerisy (scientific), 94
Clifton College (1860), 196
coal-mine owners and managers,
    238, 241–2, 247
coal-mining, 237–9, 244, 254
cosmopolitans, 32–3, 35, 37, 38, 86
counter-culture, 39–40
creationism, 189
cultural imperialism, 38, 112, 158
cultural overhead capital, 26, 32,
    103, 187, 195

decline of science, 55, 65–6, 73–5
Derby, 28, 33, 211
diffusion of science, 151–2
Doncaster, 28, 49
Dublin, 234

economic depressions, 31–2, 57–8,
    241
Edinburgh, 22–4, 27, 36–8, 151–78,
    234, 265–6
Edinburgh Philosophical
    Association, 21, 36, 153–69
Edinburgh Philosophical
    Institution (1846), 177

Edinburgh Phrenological Society
    (1820), 157
Edinburgh Review, 221
Edinburgh School of Arts, 155
Edinburgh Society for Diffusion of
    Knowledge, 158
Edinburgh Weekly Courier, 157
electricity, 95–6
English cholera, 267

Fellowship of the RS, 60, 65, 72–4,
    80–1, 135
Franklin Institute (Philadelphia),
    237

general practitioners, 260–6, 274
Gentleman's Magazine, 67
Geological and Polytechnic Society
    of the West Riding, 23, 231–56
geological societies, 234
Geological Society (London), 15,
    25, 120, 130, 231–2, 238
geology, 36–7, 231–56
Germany, 43–5
glacial theory, 239
Glasgow, 167, 265
Goole, 266

Halifax Literary and Philosophical
    Society, 223
hegemony, 17, 22, 57, 59, 83, 152
Herschel Declaration (RS), 64–6
hospitals, 93, 97, 264–5
Huddersfield, 28
Hull, 33, 36, 273

individualism, 16–17
intraprofessional conflict, 257–8,
    265, 274–5

Japan, 44

Kendal, 28

King's College (Aberdeen), 164,
    169
Kirkdale Cave, 232

Lancet, The, 63, 73, 259, 272
lecturing: financial rewards of, 98–
    100, 106, 156, 208
Leeds, 222, 241–3, 245–6, 256
Leeds Philosophical and Literary
    Society, 197, 233, 244–7
Leicester, 109
Literary and Philosophical
    Societies, 59, 94–5, 239–40,
    259
Liverpool, 30, 34, 109, 264–5, 273
Liverpool Royal Institution, 30
local boards of health (1831), 267
local politics, 34
London, 22, 35, 36–8, 79, 91–119,
    261, 270, 274
London Chemical Society, 135
London Institution (1805), 25, 94,
    97, 108, 131
London Mechanics' Institution, 97,
    103–5, 108–9
London Philosophical Society, 20,
    95, 123
Lunar Society (Birmingham), 211

Manchester, 17–18, 31, 32, 109,
    180, 242
Manchester Board of Health, 270,
    273, 277–8
Manchester College (York), 222–3,
    229
Manchester Geological Society
    (1838), 239, 244
Manchester Literary and
    Philosophical Society, 12, 15, 211
marginal men, 18–20, 32, 40–2, 52,
    122–3, 132–7, 144, 180
Mechanics' Institutes, 12, 26, 40,
    96–9, 101–5, 153, 202, 245, 256
medical education, 93, 95, 105–6

medical men, 24, 60, 65, 73, 76, 84, 92, 134–5, 187, 195–6, 204, 209, 257–78
Medico-Botanical Society (London), 106
mesmerism, 192
metropolitanism, 34–8, 257–8
Middlemarch, 258
Middleton collieries, 233
miner's lamp, 218–19, 237
Mines Act (1842), 241
mining and mineralogy, 126–32, 134, 234–5
Mining Records Office (London), 235
mining technics, 236–51
Monthly Magazine, 124, 128
Monthly Repository, 224
museum curators, 193

National Association for the Promotion of Social Science, 196
national surveys, 127
Natural History Society of Northumberland, Durham and Newcastle, 214, 234, 253
Nature, 80
Newcastle, 21, 24–5, 29, 33, 35–6, 205–30, 240, 266, 269
Newcastle Chronicle, 220
Newcastle Literary Club, 218–19
Newcastle Literary and Philosophical Society (1793), 205–30, 235
Newcastle Mechanics' Institute, 206, 217, 228
Newcastle Natural History Society, 219
Newcastle New Institution, 215–19
Newcastle Philosophical and Medical Society (1786), 209
Newcastle Philosophical Society (1775), 208

Newcastle Society of Antiquaries, 214, 219
Newcastle Unitarian Tract Society, 224
New South Wales, 193
North-Eastern Coal-field, 235
Norwich, 13, 27, 47, 126
Nottingham, 13, 47

Œconomist or Englishman's Magazine, 217
Oldham, 31
Ordnance Survey, 92
Oxbridge, 21, 59, 61, 188, 191, 195, 197, 200, 215, 231–2, 243, 262

parliament, 55–62
parliamentary grant: to the RS, 77
patents, 107
Pennine section, 244
petrology, 246
petty bourgeoisie, 154, 158–9
philanthropy, 135–6
Philosophical Club: of the RS, 74–7, 89
Philosophical Magazine, 96, 100, 124, 126
Philosophical Transactions, 64, 71, 75
phrenology, 157–9, 173–4, 192, 202, 258; *see also* Combe, G.
Phrenological Journal (1823), 157, 161, 168
platinum, 129
Plough Court (London), 124, 129, 133
Plympton, 266, 271–2
Pneumatic Institute (Clifton), 184
polite geology, 237–9, 245–51
Polytechnic Institution (London), 110
Presbyterians, 158–9
Privy Council, 262–3

professionalization, 78–9, 82–3, 111
professions, 30, 55–6, 257
Provincial Medical and Surgical Association (1832), 67
provincialism, 34–8, 109–10, 242, 257–8
public health, 261
Public Health Act (1848), 37

Quakers, 126, 132–3, 136–9, 140, 143, 254
quarantine, 261, 267, 270

rational dissent, 205–30; *see also* Unitarians
relative deprivation, 32
Requisitionists: of the RS, 63–5
Rotherham, 243
Royal Astronomical Society, 61
Royal College of Physicians, 261–3, 273
Royal College of Surgeons, 73
Royal Geological Society of Cornwall (1814), 234
Royal Institution (London), 14, 59, 92, 94, 102, 121–2, 129–32, 179
Royal Society, 21, 38, 55–90, 123
Royal Society of Edinburgh, 152, 166
Russell Institution (1808), 94

saline injections: for cholera, 271–2, 278
science lecturing, 22, 33, 91–119, 152, 160–3, 208
scientific and technical training, 43–5
scientific conservatism, 179–80, 190
scientific reform, 55
scientific societies, 84, 93–5, 96, 196
Scotland, 127, 152, 161
Scotsman, The, 155, 165

Scottish provincial science, 160–5
secrecy, 23, 189, 247
Sheffield, 29, 30, 32, 233, 242, 246, 248, 252, 256, 266
Sheffield Society for the Promotion of Useful Knowledge, 49
Shields, 266
social class, 18–19, 29–32, 39–42, 57–9, 169–71
social function of science, 41–3
social mobility, 39, 259–60
social status, 132–6
Society for Aiding the General Diffusion of Science (Edinburgh), 161–9
Society for the Diffusion of Useful Knowledge, 98, 109
Society for Preventing Accidents in Coal Mines (1813), 219
specific remedies: for cholera, 270–1
Spitalfields Soup Society, 136
Stalybridge Mechanics' Institution, 109
Statistical Society of London (1834), 238
Sunderland, 267–9
Surrey Institution (1810), 94

Thames Tunnel (1828), 94
Times, The, 78
Tyne Mercury, 221

unemployment, 32
Unitarians, 13, 19, 21, 126, 181, 185, 191, 222–5
University of Durham, 233
University of Edinburgh, 158, 164, 166, 168–9, 172, 242, 263
urban contexts, 26–34, 152–3, 179–80, 183, 207–9, 225
urban population, 27–8
urban space, 27
utilitarian science, 12–13, 22–3,

28–9, 46, 104, 107, 111, 120, 124–5, 128–9, 194, 211–13, 222, 236–44, 248

Wakefield, 233
Wakefield Literary and Philosophical Society, 233
Warrington, 28, 33, 49
Warrington Academy, 211, 222, 224
West of England Journal (1835), 194
West Yorkshire Coal Owners Association, 23, 240

Worcester, 263, 265

yellow fever, 272
York, 13, 33, 263
Yorkshire, 232–6
Yorkshire Agricultural Society (1837), 243–4
Yorkshire coal-field, 239–40, 242
Yorkshire Mining, 236–9
Yorkshire Philosophical Society (1822), 223, 232, 242, 255

Zoological Society (London), 98

Name index

Accum, F., 15, 95
Acraman, D. W., 185
Adamson, J., 219–20
Addams, R., 109, 184
Aikin, A., 125–32, 138
Aikin, C. R., 125–32, 138
Airy, G., 68
Alderson, J., 33, 215, 273
Aldini, G., 96
Alford, B., 183
Allen, W., 92, 95, 123, 125–32, 134, 136, 138
Altick, R. D., 111
Arch, A., 143
Arnold, M., 81
Ashton, T. S., 11
Atkinson, H., 220–1
Audubon, J. J., 152
Ayres, J., 273

Babbage, C., 55, 60, 63–4, 66, 69, 73, 108
Babington, W., 123, 125, 139
Bache, A. D., 237, 253
Bagehot, W., 57, 75, 78, 195

Baily, F., 92
Baines, E., 97
Bakewell, R., 184
Balfour, A., 81
Ball, J., 143
Banks, Sir J., 59, 123
Barber, J., 192
Barnes, B., 14, 18, 24, 101
Barrow, J., 68
Barry, Sir D., 263, 274
Beck, T. S., 72
Beddoes, T., 184, 217
Beeke, H., 186
Beilby, R., 216
Bell, C., 60
Bell, T., 77
Bengough, G., 194
Berman, M., 17–20, 41, 101, 122, 131, 179
Bernal, J. D., 11, 42
Bevan, B., 127
Bewick, T., 220
Biddulph, T. T., 195
Bigge, T., 214–15, 217
Bingley, R., 125, 139

Biram, B., 246, 256
Birkbeck, G., 92, 97, 101
Blackman, J., 272
Blake, W., 186
Boase, H. S., 109
Bostock, J., 14
Bowles, F. C., 184
Brande, W. T., 15, 59, 98–100
Brayley, E. W., 100, 109
Briggs, A., 28, 30, 35
Briggs, H., 240–1, 245, 248, 254
Bright, R., 185, 190
Brindley, J., 192, 203
Bromby, Revd J. E., 194
Brougham, H., 110, 153, 217, 221
Bruce, J., 220
Buckland, W., 88, 232, 243
Budd, W., 196
Buddle, J., 219, 235–7, 240, 253
Buer, M. B., 121
Bunt, T. G., 194
Burdett, F., 136
Burgess, G., 192
Burke, E., 185
Burnett, G., 100
Burns, W. L., 58
Butler, J., 183

Campbell, H., 138
Cannon, W. F., 56
Carpenter, L., 185–8, 194–6
Carpenter, M., 196
Carpenter, W. B., 76, 187, 195–6, 204
Carrick, A., 187
Carson, J., 34, 264
Castle, M. H., 185
Cave, S., 191
Chambers, R., 160
Chambers, W., 160
Chantrell, R. D., 255
Chapman, W., 209, 219
Charlesworth, E., 245, 255
Charnley, W., 210, 220

Children, J. G., 193, 203
Clanny, R., 268–9
Clark, G. T., 187
Clark, J., 225
Clay, J. T., 239, 244, 254
Cleland, J., 203
Clephan, J., 210
Cock, T., 129, 139
Cockburn, H., 153, 159
Cockerell, C. R., 187
Combe, A., 163
Combe, G., 22, 36, 154, 156–63, 165–70
Conybeare, W. D., 186, 189, 190–5, 201
Cooper, A., 15, 136
Cowper, E., 109
Cox, R., 163
Cox, R. A., 136, 139
Cox, T., 125, 139
Cramer, J. A., 191
Creagh, M., 268
Crosland, M., 45
Cumberland, G., 186, 189

Dalton, J., 15
Danby, F., 189
Daniel, T., 185
Darwin, C., 68, 190, 231
Darwin, E., 33, 211
Daubeny, C., 29, 188
Daun, R., 268, 277
Davis, J. W., 231
Davy, E., 188
Davy, H., 15, 58, 60, 98, 101–2, 128–30, 133, 135, 188, 219, 237
Dawson, Revd T., 222
De la Beche, Sir H., 75, 189, 193–4
de Morgan, A., 57, 71
Denny, H., 244, 246
Dent, E., 109
Detrosier, R., 106
Doubleday, R., 208
Douglas, J., 68

Drummond, J. L., 156
Duncan, P. B., 193
Durey, M., 22, 24

Elliotson, J., 258
Embleton, T. W., 233, 239, 241–2, 248, 252
Engels, F., 30
Estlin, J. B., 185, 187, 196
Etheridge, R., 193
Ezrahi, Y., 43

Faraday, M., 15, 64, 68, 89, 98, 102
Farish, W., 215
Ferguson, J., 208
Fitzwilliam, Earl, 239
Foote, G., 55–6
Forbes, E., 75
Fordyce, G., 140
Foster, J., 31
Fox, J., 133, 136, 143
Fox, Dr J., 143
Franck, Dr, 130
Fraser, D., 34
Fraser, W., 157, 160, 166–7
Fripp, C. B., 194
Frost, J., 94, 106
Fryer, M., 184
Fyfe, A., 176

Garnett, J., 213
Garnett, T., 15, 26, 36, 95, 102, 214, 216
Gassiot, J. P., 74
Gerschenkron, A. P., 44
George, C., 185
George, E. S., 233, 252
George, M. D., 121
Gibson, Dr, 268
Gilbert, D., 61–4, 84
Glover, R. M., 214, 220
Goodchild, J., 254
Gowing, M., 43
Graham, T., 75

Gramsci, A., 17, 59, 83
Grant, R., 100
Granville, A., 65–6, 69, 74
Gray, R., 195
Greenhow, T. M., 220
Greenough, G. B., 127, 231, 239, 243–4
Greg, W. R., 203
Gregory, W., 163–4
Grove, W., 72–7, 92
Guest, J. J., 187
Gutch, J. M., 186, 194
Gutteridge, Mr, 109

Haberfield, J. K., 185
Hamilton, Sir W., 169
Halford, Sir H., 262–7
Hallam, H., 194
Hammond, B. and Hammond, J. L., 121
Harcourt, V., 189
Harford, J. S., 22, 186–7, 203
Harrison, R., 208
Hartop, H., 239–40, 245–50
Hastings, C., 263, 276
Hays, J. N., 22
Haywood, J., 246, 256
Headlam, T. E., 220
Heaton, J. D., 245, 255
Hedley, A., 220
Hemming, J., 100
Henry, R., 267
Henry, T., 15
Henry, W., 126
Herapath, W., 192
Herschel, J. F. W., 56, 61–6, 68, 76
Higgins, B., 123
Hobsbawm, E. J., 44
Hodgson, Revd J., 219–20
Hodgson, S., 220
Hodgson, W. B., 158
Holt, H., 255
Hope, T. C., 164
Horner, L., 75, 153, 155, 157, 162–3

Horsley, J., 208
Howard, L., 15, 127–36, 143
Hughes, J., 107
Hume, Revd A., 77
Hunter, Revd J., 222
Hutton, C., 208
Hutton, W., 235
Huxley, T. H., 79–81

Inkster, I., 94–5, 122, 125, 132–3,
    135–6, 142, 259

Jackson, J., 184
Jardine, Sir H., 163
Jerrard, J. H., 195
Johnston, J. F. W., 233, 236–7, 243,
    246
Joule, J. P., 15
Joyce, F., 95
Jurin, J., 208

Kahn, J., 106
Kargon, R. H., 42
Katterfelto, G., 214
Kell, J. B., 267–9, 276
Kemble, S., 207
Kendrick, J., 33
Kenrick, J., 223
Kirwan, R., 123, 126
Knight, R., 125–32, 138
Knox, R., 94

Laird, Dr, 130
Lamb, J., 200
Lancaster, J., 133
Landes, D. S., 44
Langworthy, Mr (surgeon), 266,
    271
Lardner, D., 94, 98, 100, 109
Laslett, P., 19
Laurie, D., 276
Law, C. M., 27
Laws, C., 125, 138
Lawson, H., 133

Leach, W. E., 193
Lee, R., 72
Lee, S., 191
Lees, G., 163–4
Lewes, G. H., 168
Lindley, J., 100, 106
Lloyd, R. E., 104
Losh, J., 217–25
Lowry, W., 15, 125–32, 136, 138
Long, St J., 258
Lyell, C., 15, 56, 64, 68, 92, 190,
    231–2
Lynam, C., 138
Lyons, H., 55
Lytton, B., 69

McCann, P., 136
Macdonach, O., 43
McGrigor, Sir J., 267, 276
MacKenzie, E., 209, 218
Mackenzie, Sir G. S., 160–1
MacLeod, R. M., 21, 38
Marriott, J., 192, 203
Marshall, A., 197
Marshall, J., 216, 221
Marshall, J., 110
Marshall, J. G., 233, 252
Meller, H. E., 201
Mendelsohn, E., 56
Merton, R. K., 33, 52
Merz, J. T., 55
Mildred, S., 143
Miller, J. S., 193
Millington, J., 94, 99–100, 109
Mitchell, J., 221
Moggridge, J. H., 187
Moises, Revd E., 213, 220
Monk, J. H., 195
Moody, J., 126
Moore, D., 96
Morley, S., 196
Morrell, J. B., 36, 48
Morton, C., 233, 239–45, 247–50,
    252

Moser, C. A., 31
Moyes, H., 36, 214
Mulcaster, J., 213
Murchison, R., 87, 231
Murray, J., 22, 156
Murray, J. A., 162
Murray, W., 162–3
Mushet, D., 22, 126
Musson, A. E., 17, 121

Neale, R. S., 30
Neill, P., 163
Neve, M., 20–1, 30, 35
Newman, F. W., 195
Newman, J. H., 195
Nichol, J. P., 160–1, 163–5, 169, 177
Northampton, Marquess of, 70–7, 87

Oldham, W., 125, 139
Orange, D., 21, 24–5
O'Shaughnessy, W., 271, 277
Overton, G., 126
Owen, Richard, 87
Owen, Robert, 192, 203

Page, R., 210
Park, R. E., 40
Partington, C. F., 100
Pearsall, T. J., 246, 256
Pearson, G., 95
Peel, Sir R., 58, 62–74, 87–8, 181
Pellatt, A., 109
Percival, T., 15
Pereira, J., 100
Perkin, H., 31, 79
Pepys, W. H., 15, 95, 125–32, 138, 143
Phillips, J., 232–3, 236–40, 243–6, 249
Phillips, R., 15, 95, 109, 125–32, 135, 139
Phillips, W., 15, 126, 131, 135–6, 143

Pinney, C., 191
Playfair, J., 56
Pollard, S., 32
Porter, R. S., 36–7, 120, 126, 211
Prichard, J. C., 186, 189–90, 202
Priestley, J., 16, 48, 211

Read, D., 30, 35
Reid, D. B., 163–4, 176
Rennie, J., 100
Renwick, T., 264
Riley, H., 187, 196
Ritchie, W., 99, 109
Robinson, D., 20
Robinson, E., 17, 121, 211
Robinson, G., 268
Robison, J., 159–63, 170
Rogers, N., 133
Roget, P. M., 59, 60, 63, 72, 76, 100
Rootsey, S., 192
Roscoe, W., 191, 202
Rosse, Earl of, 76
Rotherham, J., 208–9, 226
Royle, J., 75
Rubinstein, W. D., 133
Russell, Sir W., 263, 274

Sabine, E., 76
Sanders, W., 193, 197
Sandman, P., 125–32, 139
Schofield, R. E., 211
Scoresby, W., 244–6, 250, 255
Sedgwick, A., 232, 243
Selby, J. P., 235
Shapin, S., 21–4, 25, 30, 36, 101
Sharlin, H., 43
Sharp, W., 234
Sheppard, F., 38
Simpson, J., 158, 160
Simpson, M., 239, 254
Sinclair, Sir J., 213
Skinner, Revd J., 187
Smith, A., 52
Smith, Sir J. E., 29, 188

Smith, J. G., 94
Smith, R., 184
Smith, S., 160
Smith, W., 126–7
Smout, T. C., 37
Snow, J., 266, 276
Somerville, M., 68
Sopwith, T., 235, 238–9, 243
Sorby, H. C., 245–7, 249, 255
Sorsbie, M., 210
South, Sir J., 62–3
Spence, T., 208
Spineto, Marquis, 188
Sproat, W., 267
Spurzheim, J., 188
Stancliffe, J., 36, 140, 214
Stephenson, G., 218
Stevens, W. B., 272–3, 278
Stock, J. E., 187
Stocker, R., 139
Stone, Mr: of Deptford, 109
Sturgeon, W., 15, 110
Stutchbury, S., 193–4
Sussex, Duke of, 56, 59–71, 85
Sutcliffe, R., 29
Swedenstierna, E. T. (1802–3),
    127
Swinburne, Sir J., 219–20
Symonds, J. A., 187, 194

Taylor, E. G. R., 107
Taylor, J., 109, 126, 232, 238
Teale, T. P., 255
Teed, R., 96
Thackray, A., 16–20, 31, 35, 39, 41,
    180, 206
Thelwall, J., 188
Thezard, A., 138
Thomas, W., 212
Thompson, E. P., 20, 171
Thomson, I., 208
Thomson, Dr J., 222
Thomson, J. A., 79
Thorp, W., 238–47, 249–50, 254
Thwaites, G. H. K., 196
Tilloch, A., 15, 125–32, 136, 138,
    143
Tomlinson, R., 209
Trevelyan, W. C., 235
Tupper, M., 125, 139
Turner, J., 222
Turner, Revd W., 14, 24–5, 26, 29,
    35, 126, 205–30, 235
Tyndall, J., 69

Varley, S., 95
Vaughan, R., 185
Vernon, W. J., 192
Visger, H., 194
Vivian, J. H., 234, 253

Wakley, T., 259, 263–6
Walker, A., 104
Wallis, J., 99, 104
Warburton, H., 62
Ward, W. S., 244–7, 255
Warwick, T. O., 29, 214
Watson, R. S., 214
Watson, W., 23, 126
Webster, N., 187–8
Webster, T., 189
Weindling, P., 22, 25
Weld, C. R., 80
Wellbeloved, C., 223
West. W., 244–7, 255
Wheatstone, C., 100
Whewell, W., 60, 194
White, W., 76
Wilkinson, G., 214
Williams, J., 126
Williams, L. P., 56, 66, 77
Wilson, H. H., 191
Wilson, J., 22, 126
Wilson, T., 238, 240–3, 247–8,
    254–5
Wirth, L., 41
Wollaston, W. H., 92
Woods, J., 143
Woods, S., 131

Yeats, Dr G., 126
Young, A., 129
Young, T., 15, 92, 102